Armin Weinberger

# CSCL Scripts

Armin Weinberger

# CSCL Scripts

## Effects of Social and Epistemic Scripts on Computer-Supported Collaborative Learning

VDM Verlag Dr. Müller

# Imprint

Bibliographic information by the German National Library: The German National Library lists this publication at the German National Bibliography; detailed bibliographic information is available on the Internet at http://dnb.d-nb.de.

Any brand names and product names mentioned in this book are subject to trademark, brand or patent protection and are trademarks or registered trademarks of their respective holders. The use of brand names, product names, common names, trade names, product descriptions etc. even without a particular marking in this works is in no way to be construed to mean that such names may be regarded as unrestricted in respect of trademark and brand protection legislation and could thus be used by anyone.

Cover image: www.purestockx.com

Publisher:
VDM Verlag Dr. Müller Aktiengesellschaft & Co. KG , Dudweiler Landstr. 125 a, 66123 Saarbrücken, Germany,
Phone +49 681 9100-698, Fax +49 681 9100-988,
Email: info@vdm-verlag.de

Zugl.: München, LMU, Diss., 2003

Produced in USA and UK by:
Lightning Source Inc., La Vergne, Tennessee, USA
Lightning Source UK Ltd., Milton Keynes, UK
BookSurge LLC, 5341 Dorchester Road, Suite 16, North Charleston, SC 29418, USA

ISBN: 978-3-8364-4769-0

# Contents

# Foreword

This study has been conducted in the context of the DFG funded project "Co-construction of Knowledge in Web-based Collaborative Learning: Facilitating Knowledge Convergence with Scripts and Scaffolds" by Frank Fischer and Heinz Mandl. This project was allocated in the special priority program "Net-based Knowledge Communication in Groups" initiated by Friedrich Hesse, Ulrich Hoppe, and Heinz Mandl. The study is a doctoral thesis graded 'summa cum laude' at the faculty of Psychology and Educational Sciences of the Ludwig-Maximilians-Universität (LMU) München. Many thanks go out to the PhD referees Prof. H. Mandl and co-referee Prof. J. Gerstenmaier. Their continuous counseling and advice has set me on tracks. Working with Heinz Mandl was a great privilege and inspiration that got me going in the first place. After graduation, I have continued working on CSCL (Computer-Supported Collaborative Learning) and scripts at the LMU and the KMRC, Tübingen (Knowledge Media Research Center). At both institutions, I have led the multi-disciplinary European research team CoSSICLE (Computer-Supported Scripting of Interaction in Collaborative Learning Environments), which was funded in the context of the large European Network of Excellence, Kaleidoscope. Here, some groundwork was laid on making scripts more flexible. In all this time, I have collaborated with Prof. Frank Fischer and Karsten Stegmann. We have investigated several more script types apart from the ones presented in this study as well as various CSCL phenomena, such as knowledge convergence and argumentative knowledge construction. Very special thanks for the productive time of more than 10 years of collaboration are due to Prof. Frank Fischer. In short, Frank is great as researcher, supervisor, and person. Sincere thanks go out to Karsten Stegmann for the many valuable discussions. Karsten did not only provide technical support of the learning environments under investigation here, but also contributed much in organizing and analyzing the data in the years to come. There is much support to be grateful for coming from my parents, Josef and Anna Elisabeth Weinberger, and my wife, Isabel Liberado, and our wonderful sons, Arno and Luis. My loving thanks to you!

I am now associate professor at the University of Twente, The Netherlands, working with Ton de Jong and Wouter van Joolingen on the SCY project (Science Created by You) for empowering adolescents in flexible, open-ended learning environments to engage in constructive and productive learning activities with adaptive scaffolds. So, I hope you enjoy reading this book and look out for the things to come.

Armin Weinberger

# 1 Enhancing Learning through Computer Support

Computers and the internet have enabled learners to construct knowledge in *virtual* or *online seminars* together with learning partners from all over the world, who share interests rather than time and location (Collins & Bielayczyc, 1997). Learners may engage through *computer-mediated communication* (*CMC*) in a specific form of *computer-supported collaborative learning* (*CSCL*, Koschmann, 1996). Learners choose, for example, from a course catalogue of virtual universities, such as the Virtual University of Bavaria, and access virtual seminars via the internet from any convenient place at any time. In virtual seminars, *locally distant* students may engage in *collaborative knowledge construction* at *different times*. They work on tasks together through CMC in order to acquire knowledge. In this way, learners who may have demanding schedules, such as jobholders or single parents, can access further education more easily.

Some CSCL environments may not only aim to improve access to education, but also to qualitatively change learning (e.g., *Computer Supported Intentional Learning Environment – CSILE*, Scardamalia & Bereiter, 1996; Scardamalia, Bereiter, McLean, J., & Woodruff, 1989). CSILE aims to facilitate collaborative knowledge construction as alternative to traditional classroom teaching, which is often observed as a teacher explaining a concept to more or less passive students (Scardamalia & Bereiter, 1994). CSCL-students are expected to explore complex problems by contributing their individual perspectives and resources, as well as by commenting on each others' perspectives in a shared workspace, which they can access via the internet. Students' ideas and questions are represented in a central database. This representation aims to facilitate learners to build on each others' contributions, reference each others' work, and create syntheses. Thus, a shared knowledge base of a learning community can be successively built. Several generations of students may build on, contribute to, and expand this knowledge base. Compared to students in traditional classrooms, students in some CSCL environments have been found to engage in more complex, coherent social and cognitive activities, to acquire more knowledge, and to apply knowledge from multiple perspectives (Scardamalia & Bereiter, 1994).

In contrast to these results, several studies indicate that collaborative knowledge construction is not superior to traditional classroom teaching per se, because students are rarely accustomed to constructing knowledge collaboratively (Mandl, Gruber, & Renkl, 1996). Students may therefore be even less familiar with CSCL. Furthermore, the little that students know about 'learning together' may not be transferable to CSCL. Because of this, CSCL may pose particular difficulties for students. In the following paragraph known problems of collaborative knowledge construction will be outlined. Subsequently, it will be shown how these problems may become more complicated in CSCL.

Problems of collaborative knowledge construction can be allocated to social processes as well as to cognitive processes. *Social processes* of collaborative knowledge construction may be impeded, because learners may often opt for quick consensus instead of building on each others' contributions and establishing shared conceptions of a problem (cf. Chinn & Brewer, 1993; Clark, Weinberger, Jucks, Spitulnik, & Wallace, 2003; Nastasi & Clements, 1992). For instance, learners may agree upon suggestions of the best-respected student within a learning group regardless of the quality of his/her proposals. As far as *cognitive processes* of collaborative knowledge construction are concerned, learners sometimes disregard strategies, theories or specific aspects of collaborative learning tasks (Crook, 1995; Hogan, Nastasi, & Pressley, 2000). For instance, collaborative learners rarely seem to follow an adequate sequence of problem-solving steps and often engage in behavior that has been termed as *satisficing* (Chinn, O'Donnell, & Jinks, 2000; Simon, 1955): collaborative learners often digress, oversimplify and orient themselves towards minimal requirements of collaborative learning tasks. Consequently, numerous studies indicate that the desired effects of collaborative knowledge construction often fail to emerge; learners construct inaccurate knowledge which they cannot apply and are not enabled to take multiple perspectives on a subject matter (e.g., Tudge, 1989).

How may these problems reemerge in CSCL? In addition to the demands of collaborative knowledge construction, CSCL requires new cultural techniques, e.g., handling computers as media for communication. CMC is demanding, because discussants may not immediately react to others' contributions. Especially in text-based CMC discussants may take their time to formulate their own written reactions. As a consequence, CMC discussants may have difficulties inferring where in discourse their partners are. This may cause discussants to reach different focal points and to lack in understanding one another. Accordingly, some studies found CMC to be suboptimal with respect to discourse coordination and coherence (e.g., Hiltz, Johnson, & Turoff, 1986; Kiesler & Sproull, 1992). Incoherence and coordination difficulties of CMC have been argued to impede collaborative knowledge construction (Gräsel, Fischer, Bruhn, & Mandl, 2001; Kiesler, 1992). In order to generate productive collaborative knowledge construction, neither bringing students together nor providing access to a shared database via CMC seems to suffice.

Against this problematic background of collaborative knowledge construction in CSCL environments, social and cognitive processes of collaborative knowledge construction may need *instructional support*. Instructional support in CSCL environments may, for instance, provide a structure for collaborative learners to better connect their single contributions to each other and consequently, to facilitate the construction of coherent structures of applicable knowledge within a subject matter (Scardamalia & Bereiter, 1996). Specific processes of collaborative knowledge construction can be facilitated in CSCL when students are expected to assign to their individual messages different given categories, such as 'problem,' 'what I already know,' 'new learning,' and 'my theory' (Scardamalia et al., 1989). These message types aim to foster specific collaborative task strategies. In this way, instructional support is implemented into a CSCL environment and learners are led to engage in specific discourse activities when they construct knowledge collaboratively online.

Still, there has been little empirical or theory-based research and systematic experimental variation with respect to instructional support that is oriented towards specific processes of collaborative knowledge construction realized within CSCL environments. Much is known in educational psychology about how collaborative knowledge construction can be facilitated. However, this knowledge is seldom applied in designing instructional support for CSCL environments. Although CSCL research produces fascinating technical scenarios, only a few CSCL environments build on instructional approaches developed and analyzed in educational psychology. The design of CSCL environments is typically based on what is technologically feasible or aims to assimilate CMC to face-to-face (FTF) communication, which is implicitly assumed to be an 'ideal and normal' context for collaborative knowledge construction. Therefore, some CSCL environments are designed to support discourse coordination, e.g., turn taking, in order to compensate differences between CMC and FTF. Beyond facilitating discourse coordination, more specific processes of collaborative knowledge construction may be facilitated by design of CSCL environments. Acknowledged instructional approaches based on educational psychological research explain what specific processes are essential to collaborative knowledge construction.

Probably the most influential theoretical approaches to explain, predict, and facilitate the mechanisms of collaborative knowledge construction in educational psychology derive from a *socio-cognitive perspective* (see Slavin, 1996; Webb, 1989). According to this perspective, when working in small groups, learners construct knowledge by discussing and sharing knowledge with their learning partners. Thus, collaborative knowledge construction is believed to build on specific *social* and *cognitive processes*. Students actively engage in social and cognitive processes when jointly working on learning tasks, such as applying theoretical concepts to problems, asking and answering questions, and building consensus with respect to the learning task. These specific processes may facilitate multiple perspectives on learning content and may foster applicable knowledge.

Some socio-cognitive instructional approaches aim to *facilitate processes* of collaborative knowledge construction, which falls in direct contrast to *changing conditions* of collaborative knowledge construction, e.g., by training the individual learners prior to collaboration or providing a specific incentive structure (see Slavin, 1993). Several process-oriented instructional approaches provide an external structure to facilitate specific interactions of learners (e.g., *reciprocal teaching*, Palincsar & Brown, 1984; Rosenshine & Meister, 1994). In reciprocal teaching, learners are provided with a structure for comprehending text material in small groups. This structure contains several activities in a specific sequence, which is modeled by the teacher. These activities include specific text comprehension fostering strategies that the learners are expected to apply, namely questioning, summarizing, clarifying, and predicting. First, learners read the beginning section of a text. Subsequently, one learner takes the role of the teacher. The learner's task is to ask questions on the text that should be answered by another learner. Then, the student in the teacher role tries to summarize the main ideas of the text. If necessary the learning partner completes missing subjects. Thereafter the 'teacher' identifies difficult passages of the text and tries to clear them up in collaboration with the learning partner. Finally, all learners try to predict the contents of the following text passages. Learners change teacher and learner roles for following text passages in order to assure equal involvement of all learners in collaborative knowledge construction. The adopted strategies tend to enhance learning by facilitating the learners to engage in effective processes of knowledge construction. Approaches that provide structure to collaboration have been subsumed in O'Donnell and Dansereau's (1992) *scripted cooperation* approach (cf. Derry, 1999). This approach has been well researched and has proven to effectively facilitate learning (O'Donnell, 1999). In this approach, an externally induced structure of collaborative knowledge construction processes is called a *cooperation script*. Typically, cooperation scripts combine the facilitation of social and cognitive processes. Therefore, an important, but rarely answered question is, how cooperation scripts that are oriented towards social processes, and cooperation scripts that are oriented towards cognitive processes, as well as the combination of both, facilitate collaborative knowledge construction?

The facilitation of both process dimensions should result in specific *outcomes* of collaborative knowledge construction. In contrast to traditional classroom teaching and individual studying, collaborative knowledge construction is meant to foster specific qualities of knowledge. Learners are expected to apply knowledge to a problem jointly. Collaborative knowledge construction may therefore pose a test bed for the adequacy of learners' initial problem-solving strategies. In this way, *applicable knowledge* can be regarded to as a specific learning outcome of collaborative knowledge construction. When learners establish and maintain shared conceptions of a problem in collaborative knowledge construction, they need to discuss and integrate *multiple perspectives* on a subject matter. Learner may not only acquire individual problem-solving strategies, but examine problems more closely considering alternative approaches. However, there is little research explaining to what extent instructional approaches to facilitate specific processes and specific outcomes of collaborative knowledge construction apply in CSCL environments.

In **chapter 2**, theoretical perspectives to explain and predict *processes* of collaborative knowledge construction will be discussed. In the first part of this chapter, collaborative knowledge construction will be defined and socio-cognitive theories will be portrayed. Based on ideas of transactive discourse (Teasley, 1997) and social modes of co-construction (Fischer, 2001; Fischer, Bruhn, Gräsel, & Mandl, 2002), *social processes* as they come forth in discourse will be addressed. With respect to *cognitive processes*, a model of epistemic activities (Fischer, 2001; Fischer et al., 2002) will be discussed, describing how learners work on the learning task.

In **chapter 3**, a systematic approach towards types and qualities of knowledge will be introduced. Specific *outcomes* of collaborative knowledge construction referring to applicability and perspectivity of knowledge will be classed with this systematic approach.

In **chapter 4**, an overview of how *instructional support* may be realized in CSCL environments will be discussed first. Next, instructional support will be conceptualized based on socio-cognitive approaches such as scripted cooperation (O'Donnell, 1999). Finally, approaches of CSCL research and educational psychology will be linked to build prompt-based cooperation scripts for CSCL.

In **chapter 5** the *theoretical framework* will be summarized and the *research questions* regarding processes and outcomes of CSCL supported with social and epistemic cooperation scripts will be formulated.

In **chapter 6** the *methods* of how the study was conducted and how the variables were operationalized will be introduced. Additionally, the CSCL environment of the study including the learning material will be presented.

**Chapter 7** gives an overview of the *results*. First, effects of social and epistemic cooperation scripts on social and cognitive processes of collaborative knowledge construction will be reported; subsequently, effects on learning outcomes will be presented. Then, relations between processes and outcomes of collaborative knowledge construction in CSCL environments will be reported. Finally, one case study for each of the four experimental conditions will be presented.

**Chapter 8** will discuss, interpret, and compare the findings of the empirical study to prior findings. Furthermore, the case studies will be discussed. The findings of the study will then be put into perspective concerning a number of limitations of the study. Furthermore, future research questions as well as implications for constructing CSCL environments will be outlined.

# 2 Processes of Collaborative Knowledge Construction in CSCL Environments

Thus far, *collaborative knowledge construction* has described several forms of *learning together*. This description does not include information about specific socio-cognitive processes of 'learning' or what 'together' means (cf. Dillenbourg, 1999; Mandl & Renkl, 1992).

In fields of practice, 'learning together' may have diverse meanings. When asking a student to 'learn together,' he or she might associate control and evaluation by another student in simple learning tasks, such as vocabulary testing. A teacher might associate silent group work with students taking over specific individual tasks of a major project assigned to a small group. There are many more examples of what 'learning together' might include. In accordance to Mandl and Renkl (1992), more local theories of learning together need to be applied. Thus, 'learning together' needs to be specified as *collaborative knowledge construction* for the purpose of this study.

Definitions can be built by specifying the *processes* of learning together. A general distinction has been made between cooperative and collaborative learning (Dillenbourg, Baker, Blaye, & O'Malley, 1995; Roschelle & Teasley, 1995; Webb, 1989). *Cooperative learning* is characterized by a division of learning tasks within a group of learners (Slavin, 1996). Learners work on segments of a larger task individually and compose the separate solutions afterwards. *Collaborative learning*, in contrast, involves the "mutual engagement of participants in a coordinated effort" (Roschelle & Teasley, 1995, p. 70) to work on the learning task. In collaborative learning, learners actively interact with each other on a continual basis without major interruptions by teachers. Collaborative learning has also been compared to peer tutoring or peer teaching, which is defined by the involvement of more knowledgeable peers (Chan, 2001; Cohen, 1986; Reiserer, 2003). Based on these specifications, collaborative knowledge construction can be defined:

> *Collaborative knowledge construction* refers to mutual engagement of peers of equivalent learning prerequisites in learning tasks with intertwined layers of socio-cognitive processes (cf. Dillenbourg et al., 1995) as "the result of a continued attempt to construct and maintain a shared conception of a problem" (Roschelle & Teasley, 1995, p. 70) with the goals to solve the problem and to individually acquire knowledge (Chan, 2001).

A criterion not yet addressed by the above definition is the *group size*, indicating how many learners are collaboratively constructing knowledge. Typically, learning groups should be small enough to allow any member of the group to participate in socio-cognitive processes of collaborative knowledge construction (cf. Cohen, 1994). Therefore, typical small groups consist of three to five participants (Dillenbourg, 1999). In many studies on collaborative knowledge construction, only dyads are subject to research. In dyads, however, several phenomena, which are known to affect collaborative knowledge construction, may not arise. For instance, learners may only have the opportunity for *free riding* (Kerr & Bruun, 1983) in larger groups. Learners in dyads must address each other more frequently, thus lowering the possibility of free riding.

In this chapter, socio-cognitive theories of learning associated with CSCL will be portrayed, and specific social and cognitive process characteristics of collaborative knowledge construction will be discussed in detail. Next, context factors influencing processes of collaborative knowledge construction will be introduced. Subsequently, an overview of effects of computer support on processes of collaborative knowledge construction will be given. Finally, the chapter will be summarized and the basic assumptions of socio-cognitive theories will be critically discussed.

## 2.1 Socio-Cognitive Theories of Collaborative Knowledge Construction

There are several socio-cognitive perspectives on collaborative knowledge construction. For an elaborate overview of the most recent approaches see Fischer (2002). In this work, only a sample of theoretical approaches will be discussed. This selection should include influential approaches of modern educational psychology that explain mechanisms and predict effects of collaborative knowledge construction as defined above. The theoretical approaches should have particular explanatory power for the specific research questions of this study.

The underlying assumption of socio-cognitive approaches is that processes and outcomes of collaborative knowledge construction are related. Two influential socio-cognitive approaches towards collaborative knowledge construction are (1) socio-constructivist approaches, and (2), socio-cultural approaches, which originate within the field of developmental psychology. The most quoted exponents of both approaches, however, Piaget and Vygotsky respectively, have strongly influenced modern socio-cognitive perspectives on collaborative knowledge construction (see O'Donnell & King, 1999) and have been particularly linked to CSCL (see Koschmann, 1996).

Social processes will be outlined in reference to socio-constructivist approaches and cognitive processes will be linked with socio-cultural approaches. Based on Fischer's (2001) framework, social modes of co-construction as well as epistemic activities will be discussed with reference to social and cognitive processes.

### 2.1.1 Socio-constructivist theories and social processes

Piaget has influenced a specific perspective on collaborative knowledge construction (Piaget, 1985). Individuals construct knowledge by a process called *equilibration*, which has been described as the search for logical coherence in understanding (cf. De Lisi & Goldbeck, 1999). Individuals aim to establish and maintain consistent, equilibrated models of their surroundings[*]. Knowledge construction occurs, when individuals *accommodate* cognitive structures to better describe their surroundings. Concepts are refined and/or new knowledge is constructed. Individuals may also *assimilate* a given situation into existing cognitive structures. In this way, individuals may deal with concepts without changing existing cognitive structures. Therefore, individuals may sometimes disregard or misperceive information that contradicts existing cognitive structures (Kuhn, 2001). Another option is *imitation* (cf. De Lisi & Goldbeck, 1999; Fischer, 2002). Individuals may simply pretend to adopt the perspective of another person, but do not restructure their perspective. They achieve *pseudoconsensus* rather than actual accommodation (see Brown & Palincsar, 1989). Furthermore, individuals may only make *momentary adjustments*. For instance, a student being corrected by a teacher may change his/her response, but not the misconceptions responsible for the original wrong response. Another equilibration strategy is to *respond playfully*, making light of the inconsistencies of the cognitive structures. Complex subject matters are thus often oversimplified (cf. Koschmann, Kelson, Feltovich, & Barrows, 1996).

This perspective on knowledge construction underlines the importance of inducing *perturbations* or *disequilibrium* of cognitive structures in order to foster processes of equilibration as a motivation and first prerequisite to knowledge construction. Perturbations may evolve in social interaction, when divergent views are being discussed. Neo-Piagetians therefore consider *socio-cognitive conflict* as one of the central mechanism of collaborative knowledge construction (e.g., Doise & Mugny, 1984). It has been argued that socio-cognitive conflict evolves in peer interaction rather than in traditional classroom practice (= teacher learner interaction). Peer interaction may make it more likely for learners to enter into a negotiation of multiple perspectives (Hogan et al., 2000). As learners may engage in socio-cognitive conflict during more equal or *horizontal interaction* with peers

---

[*] This basic principle, in which individuals try to explain their environment and dissolve inconsistencies, can be found in several other cognitive theories, such as attribution theory or cognitive dissonance theory.

rather than during more hierarchical or *vertical interaction* with teachers, a change of learners' cognitive structures can be initiated more easily in peer interaction than in teacher learner interaction (Hatano & Inagaki, 1991; Piaget, 1985).

Socio-cognitive conflict is not a sufficient condition for knowledge construction, but needs to be productively dissolved by coordinating divergent views in order to achieve more highly developed solutions to problems (Chan, 2001; Nastasi & Clements, 1992). Several empirical studies within the Piagetian framework have been conducted aiming to specify productive interactions (Berkowitz & Gibbs, 1983; Kruger, 1992; Kruger & Tomasello, 1986). These studies examined in particular discourse interactions of learners. Berkowitz and Gibbs (1983) focused on interactions that relate to disequilibrium from the perspective of *transactive discourse* (Dewey & Bentley, 1949). Transactive discourse is defined as "reasoning that operates on the reasoning of another" (Berkowitz & Gibbs, 1983, p. 402). Interactions may be more or less transactive. An example for a less transactive discourse is when collaborative learners juxtapose their contributions without referring to contributions of other discussants. An example for a highly transactive discourse move is integration-oriented consensus building, which is when discussants directly point out aspects of the contribution of learning partners that they themselves have not considered and build them into their own line of argumentation. Peer interaction has proven to produce more transactive discourse than teacher-learner interaction (Kruger, 1992; Kruger & Tomasello, 1986). The degree of transactivity of social interactions has been found to be positively correlated with outcomes of collaborative knowledge construction (Teasley, 1997).

The individual social interactions have not only been rated on a scale of transactivity, but have also been respected for their individual functions of collaborative knowledge construction (Fischer, 2001; Hakkarainen & Palonen, 2003). For example, integration-oriented consensus building has been regarded to as relevant for learning, not only due to its transactivity, but because integration-oriented consensus building may indicate the construction of a shared conceptual understanding. This idea of specific *social modes* has been adopted and developed by Fischer and colleagues (Fischer,

2001; Fischer et al., 2002), who identified a range of social modes of increasing transactivity.

In the following paragraphs the increasingly transactive social modes of knowledge co-construction and their distinct knowledge construction function will be introduced in more detail. These social modes are externalization, elicitation, quick consensus building, integration-oriented consensus building, and conflict-oriented consensus building.

### Externalization

Externalization (of knowledge) can be a task-related activity directed towards other group members, but is of a low level of transactivity. Nevertheless, externalization has been regarded to as a central mechanism of collaborative knowledge construction (Webb, 1989). Learners explicate their knowledge, which may make (mis-)conceptions accessible for the learners in a group. Furthermore, when externalizing, learners need to restructure knowledge to convert it into linear form. Thus, knowledge is simultaneously reorganized when is externalized (Huber, 1987). Consequently, the requirement to externalize knowledge has been connected to knowledge construction. Only the expectancy to externalize knowledge has shown to facilitate learning (King, 1989b; Renkl, 1997). Individual learners do not engage in externalization on their own, but rather, learn 'silently.' Externalization is thus mainly motivated by social situations (see Cobb, 1988).

Webb (1989) acknowledges that externalization may not be a mechanism of collaborative learning, but an indication of students with favorable individual learning prerequisites to be able to externalize knowledge on a subject. Externalization may indicate prior differences between learners rather than actual learning processes. It is commonly accepted that prior knowledge is a central factor for collaborative knowledge construction (see section 2.3; e.g., Renkl, 1997). This includes that learners with favorable individual learning prerequisites may especially profit from collaborative knowledge construction. Students with favorable prerequisites may benefit the most from collaborative knowledge construction. Externalization may have a mediating function, but cannot be made responsible for variances resulting from collaborative knowledge construction.

The uncertainty of the exact function of externalization for collaborative knowledge construction is further increased by the fact that there is disagreement on whether externalization is more beneficial for the individual who is externalizing knowledge or for the learner who listens to other group members (Renkl, 1997). In contrast to most theoretical perspectives and research on externalization (Lambiotte et al., 1988; Lambiotte et al., 1987; Nastasi & Clements, 1991; Webb, 1989), Renkl (1997) found that listening to externalization can be superior to externalizing knowledge. In his study he grouped dyads of listeners and explainers. The explainers had to externalize their knowledge on complex stochastic tasks. Listeners could question and comment on the externalization, but were asked to stay in their roles as listeners. Results show clear advantages for listeners in several transfer tasks. Renkl proposes that active externalization can be connected to considerable stress and detrimental effects on motivation and time spent on a task. Learners who externalize may have less time to reflect on a complex subject matter than listeners, but need to concentrate on coordination of communication. Therefore, beneficial effects of externalization on knowledge construction have been limited to more simple learning tasks such as text comprehension, rather than complex problem-solving (Renkl, 1997).

### Elicitation

Elicitation is a more transactive social mode than externalization. Elicitation has been described as using learning partners as a resource by asking questions (Fischer, 2001). Elicitation aims at initiating a reaction from the learning partners. In this way, elicitation has been regarded to as important for collaborative knowledge construction, because it may foster externalization. Therefore, a learning group may access a larger knowledge base by elicitation (cf. Dillenbourg et al., 1995; Fischer, 2001).

Beyond motivating externalization, elicitation may inspire further exploration when learners discover gaps of understanding (Fischer, 2001).

This may also help to discover socio-cognitive conflict (cf. Renkl, 1997). The individuals who ask questions and receive elaborate explanations may be enabled to fill in the gaps of their understanding and correct misconceptions (Webb, Jonathan, Fall, & Fall, 1995). In this respect, elicitation may be beneficial for the eliciting individual as well as for the responding group members.

Some studies showed that in more successful groups more task-related questions have been asked (e.g., King, 1994). Based on these findings, some approaches successfully foster group learning by facilitating the generation of questions, e.g., with the help of prompts (see chapter 4; King, 1999; Rosenshine, Meister, & Chapman, 1996). There are however indications that elicitation and receiving help can be detrimental when learners become dependent on this help (Webb, Ender, & Lewis, 1986). Instead of attempting to work on the learning task, students may rather seek help from teachers.

### Quick consensus building

In collaborative knowledge construction learners need to coordinate themselves to reach a common goal (typically to solve a complex problem). However, there are rarely objective criteria determining when this goal has been reached. Therefore, learners need to build consensus regarding the learning task in a process of negotiation. In Teasley's work on transactive discourse (Teasley, 1997), consensus building is more transactive than externalization and elicitation. There are different styles of reaching consensus, however, that vary in their degrees of transactivity. Quick consensus building is moderately transactive. Quick consensus building can be described as learners simply pretending to accept the contributions of their learning partners in order to continue discourse. In this way, quick consensus building may not indicate an actual change of perspective, but is rather a coordinating discourse move (Fischer, 2001).

Learners coordinate their mutual understanding in discourse by a process called *grounding* (Clark, 1992; Clark & Brennan, 1991). Grounding does not need to correspond to actual mutual understanding, but to the belief, "that the partners have understood what the contributor meant to a *criterion sufficient for current purposes*" (Clark & Schaefer, 1989, p. 262, italics added by the author). There are indications that the main 'current purpose' of learners is discourse coordination (e.g., turn taking; Weinberger, 1998) rather than "to construct and maintain a shared conception of a problem" (Roschelle & Teasley, 1995, p. 70). Thus, quick consensus building has been explicitly distinguished from other forms of consensus building relevant to collaborative knowledge construction (see Fischer, 2001).

In Weinberger's (1998) study, dyads in CMC compared to FTF dyads were asked to analyze problem cases together. Their discourse was analyzed with a coding system based on Clark's (1992) grounding approach. Although some grounding moves were found to relate to learning outcome, quick consensus building did not affect knowledge construction in CMC or in FTF. These results indicate that quick consensus building is not linearly related to outcomes of collaborative knowledge construction. Quick consensus building may instead be detrimental, when learners disregard other forms of consensus building (cf. Keefer, Zeitz, & Resnick, 2000; Leitão, 2000; Leitão & Almeida, 2000; Linn & Burbules, 1993; Nastasi & Clements, 1992). When grounding is impeded, however, e.g., in CSCL environments, quick consensus building might become a more relevant factor with respect to collaborative knowledge construction. Most suggestions to improve CSCL therefore aim to facilitate coordination and quick consensus building (Hesse, Garsoffky, & Hron, 1997).

### *Integration-oriented consensus building*

One criterion of collaborative knowledge construction is that peers may eventually establish and maintain shared conceptions of a subject matter. This can be understood as an approximate process of learners who accumulate and integrate their individual perspectives (Fischer et al., 2002; Mercer, 1995). Learners may come to better understanding by adopting each other's perspectives. Thus, integration-oriented consensus building has been regarded to as highly transactive (Berkowitz & Gibbs, 1983; Teasley, 1997).

This mode is often implicitly suggested by collaborative knowledge construction scenarios when asking learners to collaborate instead of to compete. Collaborative tasks state a common goal for learners or require learners to come to a joint solution (Littleton & Häkkinen, 1999; Pea, 1994). Learners synthesize their ideas and jointly try to make sense of a task (Nastasi & Clements, 1992).

Several aspects of integration-oriented consensus building need to be considered. Integration-oriented consensus building must be carefully distinguished from quick consensus building. As opposed to quick consensus building, integration is not mere agreement, repetition or juxtaposition of individual perspectives (Roschelle, 1996). Integration is rather characterized by the combination and sublation of positions. Integration occurs when individual learners give up initially held positions and correct themselves on the basis of peers' contributions. Learners who modify initial positions may have learned by the integration-oriented co-construction modus. An indication for integration-oriented consensus building may be that "participants show a willingness to actively revise or change their own views in response to persuasive arguments" (Keefer et al., 2000, p. 77).

In empirical studies FTF and CMC learning dyads who either were or were not supported by a task-specific structuring tool were compared in respect to integration-oriented consensus building (Fischer et al., 2002). Even though substantial differences of these groups could be found with respect to other processes and results of collaborative knowledge construction, the groups did not differ significantly regarding integration-oriented consensus building. This contradicts the importance attributed to integration-oriented consensus building for learning by the theoretical framework of transactive discussion (Teasley, 1997). This inconsistency points out that it may be difficult to distinguish between quick consensus building and integration-oriented consensus building on an operational level. In Fischer et al.'s (2002) study, quick consensus building was not analyzed. Another explanation is that the inconsistency may indicate a distortion between what is observable in discourse and what can be inferred to internal processes of knowledge construction. Therefore, discourse may be a somewhat biased indicator of knowledge construction and needs to be discussed against the background of this possible distortion.

### Conflict-oriented consensus building

The beneficial influence of conflict-oriented consensus building on learning has been considered as the most transactive social mode and has been well examined (Teasley & Roschelle, 1993). Conflict-oriented consensus building has been regarded to as a component of a central learning mechanism by Piaget (1928). Learners who are confronted with divergent perspectives may experience disequilibrium as "shock" (Piaget, 1928, p. 204). Disequilibrium may induce learners to reconsider their conceptions in order to resolve the conflict in the process of (re-)equilibration. In this way, conflict-oriented consensus building may initiate other transactive modes like externalization and integration-oriented consensus building. Reflective and constructive resolution of conflict has been related to learning (Chan, 2001; Nastasi & Clements, 1992).

Similar to elicitation, conflict-oriented consensus building is characterized by explorative behavior of the participating learners. Learners may be pushed to test multiple perspectives or find more and better arguments for their positions. This indirect effect of conflict-oriented consensus building has been confirmed by several studies (e.g., Chan, Burtis, & Bereiter, 1997).

Engaging in conflict-oriented consensus building may be socially undesirable, however. In this way, learners, who are more interested in preserving good relations to learning partners, may not benefit or may even regress in collaborative learning. In this respect, a gender effect was found. It has been shown that female students were more interested in preserving good relations rather than engaging in conflict-oriented consensus building with their same sex learning partners. As a consequence, female students were more likely to adopt inadequate conceptions in order to establish and maintain good relations and thus, were more likely to regress in same sex collaborative learning, than male students learning with male students (Palincsar & Herrenkohl, 1999; Tudge, 1992).

### 2.1.2      Socio-cultural theories and cognitive processes

In contrast to Piagetian perspectives, socio-cultural theories recognize social interaction not only as a setting eventually beneficial to individual knowledge construction, but knowledge construction as being fundamentally social and mediated by language (Vygotsky, 1978). Knowledge is constructed by *internalization* of social processes in discourse. The basic assumption of socio-cultural theories and Neo-Vygotskians is that knowledge construction is happening twice: on an internal plane and on an external, social plane. The *social speech* corresponds with *internal speech*, meaning that discourse represents cognitive processes (Vygotsky, 1986). Learners successively master cultural practices in social interaction. Prior knowledge of the individual learner is activated in discourse and becomes object to comparison. As a consequence, prior knowledge may be reorganized (cf. Cobb & Bowers, 1999). Internalization describes the process in which the knowledge that is being negotiated on a social plane becomes part of the individual cognitive structures. Vygotsky stresses that learners need to be guided by more competent partners in discourse to solve problems they could not solve alone. Thus, knowledge that they do not yet possess is discussed and may be subject to internalization.

A more competent partner guides learners in a *zone of proximal development* to apply adequate strategies to solve a problem. These task strategies may be more or less domain bound. The learners internalize these task strategies. Finally, they may individually acquire the knowledge to solve problems on their own without additional support. This idea has been re-conceptualized for collaborative knowledge construction by Neo-Vygotskian researchers.

Neo-Vygotskians emphasize that collaborative learners need to coordinate their approach towards a shared goal in order to collaboratively construct knowledge. Empirical findings on processes of collaborative knowledge construction mirror an inconsistent picture regarding the question whether only specific interactions or already active, goal-oriented participation in discourse, i.e. the amount of task-related contributions, are beneficial to learning outcome.

According to Cohen's (1994) analysis of several studies, task-related activities affect learning outcome only if learners collaborate on tasks that are *complex* and *resource interdependent*. Complex tasks can be characterized by the fact that they do not provide one single, correct solution or problem-solving procedure. Resource interdependent tasks cannot be solved by one learner alone. An effect of mere participation in collaborative work on tasks with one defined correct solution that may also be solved by individual learners cannot be found (Webb, 1991).

In more recent studies, several different task-related or *epistemic activities* have been identified in collaborative knowledge construction. An epistemic dimension refers to the tasks learners are confronted with, e.g., categorizing or defining new concepts (Fischer, 2001; Fischer et al., 2002). Therefore, epistemic activities may describe how learners work on the task in more detail and in a more systematic way within a specific domain. Pontecorvo and Girardet (1993) developed a system of domain- and content-specific epistemic activities to analyze how learners explain problem cases, e.g., defining categories and words. In problem-based learning several phases and activities have been identified (Barrows & Tamblyn, 1980; De Grave, Boshuizen, & Schmidt, 1996; Schmidt, 1983), e.g., identification of various problems and linking theoretical concepts to case information (see also Stonewater, 1980; Voss, Greene, Post, & Penner, 1983; Vye et al., 1997). Hakkarainen and Palonen (2003) classified the objects of inquiry as cognitive activities of learners. They analyzed whether learners referred to linguistic form, research questions, research methods, information, explanation or other, unspecified objects when constructing knowledge together, and found those learners who were more explanation-oriented to acquire knowledge better.

Fischer et al. (2002) differentiated in collaborative knowledge construction with complex problem cases to what extent learners relate to case information, to what extent learners relate to theoretical concepts, and to what extent learners construct relations between theoretical concepts and case information. More successful learners have often left the concrete level of case information in these studies behind, and related to theoretical concepts instead. This may indicate that there are different qualities of how learners relate to the task. Epistemic activities, which have been applied by Fischer et al. (2002) for an analysis of collaborative learning with complex problem cases, will next be discussed in more detail. These epistemic activities include *construction of problem space*, *construction of conceptual space*, and *construction of relations between conceptual and problem space*. Additionally, learners may also engage in *non-epistemic activities*. Whereas all epistemic activities are on-task, non-epistemic activities can be described as off-task behavior. At this point, there is little research on how knowledge is collaboratively applied in discourse (Fischer et al., 2002; Teasley & Roschelle, 1993). From a Vygotskian perspective the tasks learners collaboratively complete should relate to what tasks learners can accomplish individually after collaboration. Learners who are able to apply knowledge when they are supported in collaboration should also be able to transfer and apply this knowledge in subsequent tasks individually.

### Construction of problem space

In order to solve problems, learners need to acquire an understanding of the problem. Therefore, learners select, evaluate and relate individual components of problem case information. On the one hand, the importance of the construction of the problem space has been outlined (Fischer et al., 2002): this epistemic activity aims at understanding the problem, which is prerequisite to successfully solve a complex problem. The construction of the problem space includes the evaluation of particularities of a problem and therefore may also facilitate the transfer of knowledge to different problem spaces (cf. Greeno, 1998). On the other hand, a focus on the concrete level of problem case information may hint to an engagement of learners in the learning task on a low level (Salomon & Perkins, 1998). In this way, learners may retell rather than interpret a problem. Accordingly it has been shown that discourse beyond a concrete level of the problem space may reflect better strategies in learning scenarios based on complex problems (Fischer et al., 2002; Hogan et al., 2000).

### Construction of conceptual space

In order to solve problems on the grounds of theoretical concepts, learners need to share the understanding of a theory. To this end, learners construct relations between individual theoretical terms or principles. The construction of conceptual space also implies to distinguish concepts from each other. During this process, concepts are being defined and categorized by the learners (cf. Pontecorvo & Girardet, 1993). Within collaborative knowledge construction environments the construction of conceptual space has been argued to be essential for successful problem-solving (De Grave et al., 1996).

### Construction of relations between conceptual and problem space

The construction of relations between conceptual and problem space has been regarded to as the main task in knowledge construction based on problem-solving (De Grave et al., 1996). The individual relations between concepts and problems that learners construct can indicate to what extent learners are able to apply knowledge adequately, as well as how learners approach a problem in detail. Therefore, relations between conceptual and problem space can indicate, which concepts learners resort to in order to solve the problem. With respect to a Vygotskian perspective, the collaborative application of theoretical concepts to problem space may indicate the internalization of these relations between conceptual and problem space. In other words, learners who apply theoretical concepts to problems collaboratively may be able to transfer this knowledge to future problem cases, as well as apply theoretical concepts individually.

### Non-epistemic activities

Learners may not only deal with the contents of the learning task, but may also digress off-topic. For instance, learners may choose to talk about themselves or coordinate off-task aspects of their environment, e.g., asking learning partners how much time is left. A high frequency of non-epistemic activities has been connected to detrimental effects on knowledge construction as cognitive resources may be occupied and learners can be distracted (cf. Cannon-Bowers & Salas, 1998; Cannon-Bowers & Salas, 2001; Fischer, Bruhn, Gräsel, & Mandl, 2000).

## 2.2    Relations between Processes and Outcomes of Collaborative Knowledge Construction

Socio-cognitive perspectives on collaborative knowledge construction are based on the assumption that processes influence outcomes of collaborative knowledge construction, e.g., the individual acquisition of knowledge. Therefore, this relation between processes and outcomes of collaborative knowledge construction has been subject to research from socio-cognitive perspectives on collaborative knowledge construction. Some approaches have considered the frequency of individual phenomena in discourse of learners, e.g., giving explanations (Webb, 1992). Other approaches have analyzed discourse on a global scale of transactivity (Teasley, 1997) or of being epistemic or not (Cohen, 1994). In any way, utterances of individual learners have been successfully linked to individual knowledge acquisition.

Some approaches point out, however, that relations between processes and outcomes may not be linear, but reciprocal (e.g., Bandura, 1986), implying for instance, that learners who ask twice as many questions do not necessarily learn twice as much. This concludes that some discourse phenomena may not relate linearly to learning outcomes, but can still pose an essential component of discourse with specific functions, e.g., externalization that initializes discourse (Bruhn, 2000). The group mean of any discourse phenomenon may poorly relate to learning outcome (cf. Sellin, 1990). Furthermore, the distinct discourse phenomena may interact with each other. Therefore, individual utterances as well as sequences and structures of discourse may need to be analyzed (Chan, 2001; Chinn et al., 2000). Consequently, instead of counting only individual discourse phenomena, discourse types of collaborative knowledge construction may be identified. Some studies, for instance, report substantial relations between argumentative structures in learners' discourse and individual knowledge acquisition (Anderson, Chinn, Chang, Waggoner, & Yi, 1997; Chinn et al., 2000; Keefer et al., 2000; Leitão, 2000).

These more recent approaches are typically based on a variety of methodologies due to limitations of the individual qualitative and quantitative research traditions (Chi, 1997). Based on the works of Walton and Krabbe (1995), Keefer et al. (2000) apply a graphical coding analysis and identify critical discussion as the most productive type of discussion for collaborative knowledge construction, because it may facilitate understanding and accommodation of divergent viewpoints. Critical discussion is characterized by learners probing into, revealing, anticipating, and challenging the perspectives of the co-discussants. Critical discussion may help learners "to discover not only their opponents underlying positions, but their own as well" (Walton & Krabbe, 1995, p. 188). Other discussion types include "consensus dialogue," in which all learners agree on the first claim presented. Keefer et al. (2000) also discuss "eristic dialogue," in which learners attack partners' positions without concessions to their own points of view. In response to the question, what kind of processes – social and cognitive – are relevant for collaborative knowledge construction, it is yet unclear whether there are patterns concerning a social and a cognitive dimension of discourse that relate to knowledge construction, or if contributions of individual learners predict knowledge acquisition best.

## 2.3 Context Factors Influencing Processes of Collaborative Knowledge Construction

What are the factors that influence the processes of collaborative knowledge construction? There has been ample research on context factors that are believed to affect how learners collaboratively construct knowledge. These include the organizational background, incentive structure, learning task, individual learning prerequisites, and cooperation scripts (cf. Renkl & Mandl, 1995). In this section, these context factors will be briefly introduced.

### *Organizational background*

With regard to organizational background, the pivotal concern is, to what extent collaborative knowledge construction is an accepted learning form in an educational institution. It has been shown that the individual endeavors of teachers to encourage students to practice collaborative knowledge construction rarely lead to the desired outcomes (McLaughlin, 1976). Instead, collaborative knowledge construction needs to be embedded into school routine, and needs to be supported by school administration as well as by the students. Learners may need to systematically experience the benefits of collaborative knowledge construction in order to better accept and to develop the prerequisite competencies of collaborative knowledge construction (Renkl, Gruber, & Mandl, 1996; Renkl & Mandl, 1995). Furthermore, organizational background also includes aspects such as group size, room size, time frames, and the financial conditions which collaborative knowledge construction may afford (Huber, 1999).

### *Incentive structure*

Slavin (1993) emphasized group reward and individual responsibility as necessary conditions for collaborative knowledge construction. He argues that learners in groups need to be interdependent with respect to achieving shared rewards. Simultaneously, individual contributions to group performance need to be salient in this model, meaning that the performance of the individual group member needs to be identifiable. Cohen (1994) restrains the validity of this approach to training tasks, in contrast to complex problem-solving tasks. The extent to which the typical incentive structure in educational institutions facilitates motivation to engage in collaborative knowledge construction has been discussed. Evaluation of students' performances is typically based on individual tests on conceptual knowledge, which can also be individually constructed. Knowing that collaborative knowledge construction may be demanding and may require much effort, learners may wonder why they should learn together. In this way, *incentive structure* has been argued to define the goal motivation of learners. In school contexts, for instance, the motivation of students in learning environments may be different from those intended by teachers. Students may focus on certification, i.e. passing the exam, or may try to impress peers. These goal orientations may conflict with goals of teachers to facilitate conceptual understanding. Similarly, goal orientation of participants in experimental learning environments may differ from that of learning, e.g., receiving a certificate, earning money (cf. Ng & Bereiter, 1991). To these ends, learners may not aim to be dependent on learning partners and establish shared goals and conceptions (see also Johnson & Johnson, 1992).

### *Learning task*

The *collaborative learning task* is another context factor for effective collaborative knowledge construction. It has been argued that several task criteria need to be met for social and cognitive processes of collaborative knowledge construction to evolve (cf. Cohen, 1994). First of all, the task should be *complex*, so that learners need to join resources (e.g., effort, competencies, material) to solve it (*resource interdependence*). Cohen (1994) argues that a major part of research on collaborative knowledge construction confronts collaborative learners with simple tasks with only one correct solution. These tasks, however, can not be referred to as actual group tasks. Learners may not need to construct and maintain shared conceptions in order to solve simple tasks. Furthermore, the task should be *motivating* in order to reduce satisficing (cf. Renkl & Mandl, 1995). The Cognition and

Technology Group at Vanderbilt (1992), for instance, has developed a video format to present complex problems as learning tasks. In these videos, learners are requested to help a 'hero' saving a wounded eagle by calculating distances, gas consumption etc. in order to foster the motivation of learners.

### Individual prerequisites

Several individual learning prerequisites have been discussed as important factors for collaborative knowledge construction. These individual learning prerequisites may be categorized as cognitive (prior knowledge and learning strategies) or emotional and motivational (social anxiety, uncertainty orientation, interest). With respect to CSCL, computer-specific attitudes have also been discussed.

*Prior knowledge* is a central prerequisite to collaborative knowledge construction (e.g., Stark & Mandl, 2002). According to constructivist approaches and perspectives of cognitive elaboration, new knowledge needs to be connected to prior knowledge. The more prior knowledge, the more potential connection points for new knowledge are available (Dochy, 1992). This influence of prior knowledge on learning outcome can be particularly found in learning environments that are based on complex learning tasks (Weinert, 1994). Furthermore, prior knowledge has been reported to interact with facilitation of collaborative knowledge construction (see chapter 4; O'Donnell & Dansereau, 2000). Some facilitation tools may be particularly useful for learners with high prior knowledge. Some other facilitation tools may support learners with low prior knowledge, but will impede progress of learners with high prior knowledge (Cohen, 1994; O'Donnell & Dansereau, 2000).

*Learning strategies*, which describe how learners deal intentionally with learning material, may influence how well learners understand theoretical texts (Wild & Schiefele, 1994). Learners who read through learning texts critically, for instance, may have advantages over learners with inadequate learning strategies when applying theoretical concepts to a problem.

*Social anxiety* is a condition, which results from the perception of a socially demanding situation that is interpreted as a threat, typically to self value. Particularly for collaborative knowledge construction, social anxiety correlates negatively with performance (cf. Seipp & Schwarzer, 1991). It has been argued that anxiety also influences the application of learning strategies (e.g., Naveh-Benjamin, 1991). The cognitive capacity of anxious learners seems to be consumed by the anxiety. Anxious learners may have difficulties concentrating on the learning task or applying adequate task strategies. Learning groups in discourse without instructional support reported higher levels of social anxiety than dyads provided with cooperation scripts (O'Donnell, Dansereau, Hall, & Rocklin, 1987).

*Uncertainty orientation* has been found to be an individual characteristic relevant to motivation, learning strategies, and learning outcome in collaborative knowledge construction environments (Huber, Sorrentino, Davidson, Eppler, & Roth, 1992; Sorrentino, Short, & Raynor, 1984; Stark, Gruber, Renkl, & Mandl, 1997). This construct explains and predicts to what extent individuals seek additional information, explore inconsistencies, or try to get along with available knowledge and avoid conflicting information (Dalbert, 1996; Huber et al., 1992). The socio-constructivist theories on collaborative knowledge construction, which are outlined in this work, strongly build on the idea that learners engage in productive equilibration of socio-cognitive conflicts. Thus, uncertainty orientation has been found to positively relate to outcomes of collaborative knowledge construction (Huber et al., 1992; Stark et al., 1997). Certainty oriented learners, in contrast, aim to avoid socio-cognitive conflicts, which is detrimental for individual knowledge acquisition in collaborative knowledge construction environments. Therefore, certainty oriented learners often aim to avoid collaborative knowledge construction environments altogether. These learners prefer individual learning environments.

*Interest* in the subject matter of the learning environment can be regarded to as an individual characteristic, which strongly influences learning outcomes (Krapp, 1999). Learners with more interest in the subject matter may more likely invest more effort to approach problems from multiple perspectives. Learners with little interest, however, may show satisficing and aim to accomplish minimal requirements of a collaborative learning task.

*Computer-specific attitudes* have been argued to be individual characteristics strongly influencing CSCL (Richter, Naumann, & Groeben, 2001). Richter et al. (2001) have differentiated several dimensions of attitudes towards computers, for instance, perceiving computers as a threat to society. These attitudes towards computers have shown to influence the motivation to learn in CSCL environments (Richter et al., 2001).

*Cooperation scripts*, too, have been denounced a context factor for collaborative knowledge construction (O'Donnell & Dansereau, 1992; Renkl & Mandl, 1995). The activities of collaborative learners may be guided by cooperation scripts. Cooperation scripts provide a structure to collaborative knowledge construction by *specifying*, *sequencing*, and *assigning roles or activities* to learners. A number of instructional approaches have been designed to foster collaborative knowledge construction based on the idea of cooperation scripts (O'Donnell & Dansereau, 1992). For instance, in the *jigsaw method*, learners are assigned an expert role for one specific question regarding the subject matter that is to be learned (Aronson, Blaney, Stephan, Silkes, & Snapp, 1978). Then, learners build expert groups, in which they discuss this one specific question. Subsequently, learners of the various expert groups build basis groups in a way that for any question on the subject matter one expert is present. In these basis groups, learners are expected to convey their specific expert knowledge. By specifying, sequencing, and assigning roles and activities, cooperation scripts do not only suggest activities to learners, which are believed to facilitate knowledge construction in turn. They also assure that learners are equally involved in establishing and maintaining shared conceptions and can approach a problem from multiple perspectives. This is typically achieved with cooperation scripts that require learners to change roles during collaboration (e.g., *reciprocal teaching*, Palincsar & Brown, 1984). The cooperation script approach is the issue of this work and will be discussed in more detail in chapter 4.

Another context factor, which has not yet been discussed by Renkl and Mandl (1995), are the *tools* learners use to communicate and work on a joint learning task. CSCL based on CMC as a context factor for processes of collaborative knowledge construction poses another major issue of this work and will be discussed next in more detail.

## 2.4 CSCL Technology and its Influence on Collaborative Knowledge Construction

There is a wide range of how computers may influence collaborative knowledge construction. The computer may visualize collaborative knowledge construction, provide learning material, and mediate communication of learners (Mandl & Fischer, 2002; Mandl & Fischer, 2000). A range of techniques has been developed to mediate communication via the computer (e.g., e-mail, chat, video conferencing). Therefore, there are many forms of computer-based media with distinct qualities (Weinberger & Mandl, 2003). Most CSCL environments are based on text-based CMC on digital discussion boards. Text-based CMC is technically feasible with standard computer equipment and low-bandwidth internet access (cf. Weinberger & Mandl, 2003). Furthermore, it has been argued to simultaneously support permanent visualization of group processes and, consequently, facilitate more reflective discourse (Straus & McGrath, 1994). In this section text-based CMC will be introduced. Subsequently, empirical findings on the influence of text-based CMC on social and cognitive processes of collaborative knowledge construction in CSCL environments will be discussed.

### 2.4.1 Technology for CSCL environments: Text-based CMC in web-based discussion boards

The most disseminated form of CMC is text-based communication, which will be presented next (see Weinberger and Mandl (2003) for more detailed descriptions of other forms of CMC). Text-based CMC means that discussants type what they want to communicate, send, receive, and read messages onscreen (Döring, 1997b). Text-based CMC can be asynchronous. In *asynchronous CMC*, discussants are not expected to interact at the same time, but any non-technical delay between the individual activities may take place. This means that messages can be recorded and discussants can respond at any later, convenient time. In most current CSCL environments, learners communicate via text-based, asynchronous discussion boards. In *discussion boards*, messages are recorded on a central database and typically represented in discussion threads. These threads start with one particular message that is indicated in a message overview by its title, the author, and the date of entry. Any response to a message is graphically connected to a message that initiates a discussion by a line or 'thread' and indented. Thus, an increasingly indented discussion thread is built in which the discussants are supposed to continue the specific subject which was initialized with the very first message. New subjects are meant to be set off with a new discussion thread. Discussion boards have been realized as newsgroups with distinct computer servers and software. More recently, discussion boards have been also implemented into WWW (World Wide Web) applications, which can be referred to as *web-based discussion boards* and which can be accessed with standard internet browsers. Compared to FTF discourse, text-based CMC differs in some respects. FTF-communication deploys in a linear fashion. In web-based discussion boards, however, CMC discussants may enter discourse at various points of a discussion thread. CMC discussants may also enter asynchronous discourse at any time. Thus, they can take any time to formulate messages. Another important technological aspect of text-based CMC is the *reduction of social context cues*. CMC discussants may remain anonymous to a certain extent (cf. Weinberger & Mandl, 2003). Participants may use pseudonyms and may not possess any further audio-visual information about each other such as dialect, gender, skin color etc. Most approaches to analyze effects of text-based CMC on sharing knowledge are based on this filter effect of the medium (see below).

## 2.4.2 Influence of CMC on social and cognitive processes of collaborative knowledge construction

There has been ample research and discussion about if and how computer-mediated communication may influence processes of collaborative problem-solving and knowledge construction. In the early 1990's it was debated whether or not technology would at all influence how knowledge is constructed. One position was that there is no media influence on collaborative knowledge construction (cf. Clark, 1994). It has been argued that the medium is a mere vehicle and does not turn information into knowledge. Although computer-based media shows some potential for learning, e.g., visualizing collaborative knowledge construction, this potential is not exclusive to computer-based media. In this way, results of CMC studies cannot be attributed to the medium. In contrast to this position, CMC has been understood as a cultural practice, which de facto has impact on the way knowledge is constructed and communicated. The medium provides a specific context for discussants and consequently needs to be considered when analyzing collaborative knowledge construction (cf. Gerstenmaier & Mandl, 2001). Therefore, CSCL research needs to aim towards analyzing and facilitating this cultural practice with computer-based media (cf. Jonassen, Campbell, & Davidson, 1994).

Many studies show differences between CMC and FTF-communication. Discussants may communicate differently, because they may be more anonymous, because they may have more time to formulate their contributions, or because they have to type what they want to communicate, etc. The results of this research have been rather inconsistent, however (cf. Fabos & Young, 1999). These inconsistencies may be explained by different, media- and user-centered approaches towards analyzing CMC (cf. Weinberger & Lerche, 2001).

Some researchers have taken a *media-centered approach* in order to explain different results between CMC and FTF groups. This approach emphasizes the idea that in CMC the communication channel is reduced and social context cues are filtered out. This *channel reduction*, which is particularly strong in text-based CMC, has been argued to lead to a range of effects on communication. CMC discussants are less likely to recognize each other's social status. Therefore, CMC may reduce inhibitions caused by status differences to avoid *conflicts* (Kiesler, Siegel, & McGuire, 1984). Furthermore, the social context cues, which are believed to be filtered out in the reduced CMC channel, usually support the coordination of FTF discussants (e.g., turn taking). Due to the resulting coordination difficulties, text-based CMC is often characterized by less frequent turn taking, longer individual messages, and incoherent discourse (Hesse et al., 1997; Quinn, Mehan, Levin, & Black, 1983). Therefore, text-based CMC groups take more time to come to conclusions and have been considered to be *less productive* than FTF groups in tasks that demand a high frequency of turn taking, e.g., solving complex problems (Straus & McGrath, 1994).

Only in idea-generating tasks have text-based CMC groups performed equally well as FTF groups (Dubrovsky, Kiesler, & Sethna, 1991). This effect has been ascribed to the possibility of giving input simultaneously in text-based CMC. In contrast, members of FTF groups may mutually block the production of ideas as each discussant is expected to wait for his or her turn (*production-blocking effect*, Fischer, 2001). Channel reduction has also been associated with some potentially beneficial effects of text-based CMC. In comparison to FTF-communication, CMC has also been found to facilitate epistemic activities and conflict-oriented consensus building, which has been positively related to collaborative knowledge construction (Diehl & Ziegler, 2000; Kiesler, 1992; Rice, 1984; Riel, 1996; Woodruff, 1995).

*User-centered approaches* criticize the emphasis of technical characteristics of CMC and its effects on intra- and interpersonal conditions. Thus, findings on channel reduction of CMC have been put into perspective by user-centered research that considered *time* as an important constraint in text-based CMC. Text-based CMC groups may perform equally well as FTF groups, but may require more time due to the typing lag (Walther, 1996). Groups that communicate in a computer-mediated way for longer periods of time have often developed a discourse comparable to FTF groups (Spears, Lea, & Lee, 1990; Walther, 1992). These results indicate that any former channel reduction research is particularly valid for any anonymous ad-hoc groups, which interact for short periods of time only. Virtual groups in real world settings typically start out as anonymous groups, and some online learning scenarios may be based on short-term cooperation. However, when online collaboration lasts over longer periods of time, users may compensate the channel reduction effects of text-based CMC. This means that social context cues may not be filtered out completely, but the user may evaluate diction, provide personal background information (e.g., homepages), and simulate social context cues in a text-based manner (Döring, 1999). For instance, discussants may enrich text-based CMC with 'emoticons' or 'smileys' (e.g., :-)), comic language (e.g., *grin*), abbreviations specific to text-based CMC (e.g., ROTFL = roll on the floor laughing), or TYPING IN CAPITAL LETTERS, which is considered to be screaming.

Moreover, CSCL environments may be designed for specific purposes, such as compensating barriers and fostering potentials of CMC for collaborative knowledge construction (cf. Hesse et al., 1997). For example, CSCL environments may be designed for communities of learners that introduce themselves in detail on individual homepages (cf. Reinmann-Rothmeier & Mandl, 2001). Therefore, CMC participants may be provided with even more personal information about other group members than groups that meet face-to-face. This modifiability of CSCL environments may result in a range of possible effects on group processes (cf. Lea & Spears, 1991). In this rationale, technical aspects may not be regarded to as definite characteristics of CMC, but modifiable for educational purposes in specific CSCL environments. Thus, effects on social and cognitive processes may strongly depend on experience of learners to use text-based CMC and how interfaces are designed.

## 2.5  Conclusions and Limitations: Socio-Cognitive Processes in Learners' Discourse

The socio-cognitive theories of learning indicate that collaborative knowledge construction builds on both social and cognitive processes. How learners interact with each other describes the social dimension of collabora-tive knowledge construction. The studies on transactive discourse point out that learners need to relate the reasoning of each other in order to construct knowledge successfully. Simultaneously, more or less transactive interac-tions may have specific functions in collaborative knowledge construction. How learners relate to and deal with the task appears to be a second cogni-tive dimension of collaborative knowledge construction. Cohen's (1994) analysis of several studies underlines frequency of epistemic activities as an indicator for collaborative knowledge construction. More recent studies have distinguished several epistemic activities with a differentiated effect on the learning outcome. Based on the socio-cognitive theories of learning out-lined above, social and cognitive processes of collaborative knowledge con-struction have been discussed with reference to Fischer's (2001; Fischer et al., 2002) approach on *social modes of co-construction* and *epistemic activi-ties*.

CSCL environments have been argued to affect social and cognitive processes of collaborative knowledge construction. With respect to social processes it has been found that CMC learners appear to engage more easily in conflict-oriented consensus building, but generally refer less to each other's contributions compared to FTF learners. Regarding cognitive proc-esses, CMC has been found to facilitate epistemic activities and multiple approaches towards the task compared to FTF-communication. Further-more, CSCL environments may be designed to foster specific social and cognitive processes. This rationale has been applied successfully in various studies (see Scardamalia & Bereiter, 1996). A specifically designed inter-face suggests specific discourse moves. The discourse of learners has been argued to reflect socio-cognitive processes of collaborative knowledge con-struction and thus an improved discourse may lead to improved collabora-tive knowledge construction (Leitão, 2000; Vygotsky, 1986).

Several phenomena may be observable that constitute discourse of learners, but only specific interactions and discussion types may be account-able to collaborative knowledge construction. Not just any kind of talk may indicate knowledge construction. Precisely how discourse may correspond with actual learning processes has been only poorly examined at this time. It is plausible to assume that discourse reflects some of the social and cognitive processes. However, our understanding of the relation between language and thought is vague. Both 'social and internal speech' (Vygotsky, 1986) may not be identical, but rather induce each other. Ideas that may emerge in discourse, but are not immediately followed up, may resurface at a later point in discourse. Thus, ideas may not be lost, but hard to trace. This disappearance and re-emergence of ideas between the plane of social and internal speech has been described as a 'dolphin' effect by Mercer (1994, as cited in Scanlon, Issroff, & Murphy, 1999). In comparison to other forms of assessment methods, like 'thinking aloud' (Mandl & Friedrich, 1992), discourse may offer a more 'natural' access to processes of collaborative knowledge construction, with both social and cognitive processes being mutually dependent.

Apart from methodological questions, other approaches to collaborative knowledge construction exist which stress modeling and imitation of behavior as important factors for collaborative knowledge construction (Bandura, 1986). Learners who do not participate actively in learning discourse may profit in collaborative scenarios by mere observation of a more capable peer. These learners may resolve socio-cognitive conflicts not during collaborative phases, but at a later time (cf. Doise & Mugny, 1984; Howe & Tolmie, 1999; Littleton & Häkkinen, 1999). In this way, individuals may continue to defend their point of view in front of peers, but may come to more appropriate perspectives and apply them individually following collaborative phases. Furthermore, some more advanced learners may have a 'secret master plan,' which means they have a functional model of the problem that they do not want to share (Fischer & Mandl, 2000). Those learners may not recognize the need to engage in collaborative activities, because they can solve the problem on their own. It could be argued that when learners dispose of a secret master plan, learning outcomes may not correspond to the social and cognitive processes of collaborative knowledge construction at all. Therefore, various approaches to analyze collaborative knowledge construction need to complement each other. In addition to the analysis of individual discourse phenomena, discourse structures may need to be analyzed as well (Chinn et al., 2000). Furthermore, qualitative and quantitative analysis methods may need to complement each other (Chi, 1997).

# 3    Outcomes of Collaborative Knowledge Construction

Most studies of educational psychology aim to improve learning outcomes, which typically means to foster knowledge acquisition. Recent evaluations of educational institutions have posed new requirements towards improving learning outcomes. Not only should the amount of conceptual knowledge learned be fostered, but also specific qualities of knowledge. In traditional classroom teaching, the knowledge students acquire is often of low quality. Basic problems of knowledge quality are, that the acquired knowledge is often based on misconceptions, is often compartmentalized, and remains inert (Bransford, Franks, Vye, & Sherwood, 1989; Mandl, Gruber, & Renkl, 1993, 1994b). Learners are not enabled to approach problems from multiple perspectives and solve problems based on theory, but are instead requested to memorize individual concepts. Different ideas of what can be expected as learning outcome have been connected to collaborative knowledge construction scenarios. Collaborative construction of knowledge, which is characterized by inquiry and discussion of complex problem cases, aims in particular to facilitate applicable knowledge and multiple perspectives (Dochy, Segers, van den Bossche, & Gijbels, 2003; Fischer et al., 2002; Renkl, 1997; Spiro, Feltovich, Coulson, & Anderson, 1989).

In this chapter, an overview of several types and qualities of knowledge will be given and subsequently, learning outcomes of collaborative knowledge construction will be specified. It will be argued that in accordance with a knowledge-in-use metaphor (see De Jong & Fergusson-Hessler, 1996), applicable knowledge can be regarded to as a co-construct and as individually acquired knowledge (Fischer, 2001). Furthermore, focused and multi-perspective applicable knowledge will be allocated in De Jong and Fergusson-Hessler's (1996) matrix of types and qualities of knowledge, and it will be discussed as outcome specific to collaborative knowledge construction. Finally, the outcomes of collaborative knowledge construction will be summarized and the knowledge-in-use metaphor will be critically discussed.


## 3.1      Types and Qualities of Knowledge

Many knowledge constructs have so far been conceptualized. The high amount of knowledge types that are being discussed may indicate the range of functions of what we know. The types of knowledge describe what content knowledge is about, e.g., knowledge about conceptual space, and what function knowledge fulfills regarding a specific task. Additionally, different qualities of knowledge have been discussed. In this way, knowledge has been characterized, for instance, as being superficial or deep, compartmentalized or coherent, inert or applicable. It can be assumed that knowledge qualities help to distinguish expert from novice knowledge. Knowledge qualities may, therefore, provide an orientation on deficits of learners' knowledge and guide instructional approaches towards reducing these deficits. In order to systematize the various perspectives upon knowledge, De Jong and Fergusson-Hessler (1996) introduced a matrix considering both the type of knowledge and the quality of knowledge, with respect to learning tasks that involve problem-solving. This matrix of types and qualities of knowledge will be introduced in this section.


### 3.1.1      Types of knowledge

De Jong and Fergusson-Hessler (1996) assume the epistemic perspective of knowledge-in-use, meaning that "task performance forms the basis for the identification of relevant aspects of knowledge" (p. 105). In this perspective, the function of the various types of knowledge for a problem-solving task is emphasized. Thus, there may be some links between De Jong and Fergusson-Hessler's (1996) types of knowledge for problem-solving and Fischer's et al. (2002) epistemic activities. De Jong and Fergusson-Hessler (1996) distinguish situational knowledge, conceptual knowledge, procedural knowledge, and strategic knowledge. In the following paragraphs these types of knowledge and their corresponding epistemic activity will be summed up briefly. With respect to strategic knowledge the parallels to scripts sensu Schank and Abelson (1977) will be discussed.

### Situational knowledge

"*Situational knowledge* is knowledge about situations as they typically appear in a particular domain" (De Jong & Fergusson-Hessler, 1996, p. 106). Corresponding to the construction of problem space, situational knowledge may indicate that learners have acquired a representation of the problem and are able to abstract problem features. Individuals who have acquired situational knowledge may therefore have understood characteristics and categories of a range of problems within a domain.

### Conceptual knowledge

"*Conceptual knowledge* is static knowledge about facts, concepts, and principles that apply within a certain domain" (De Jong & Fergusson-Hessler, 1996, p. 107). It can be argued that the epistemic activity of the construction of conceptual space corresponds with conceptual knowledge-in-use. Thus, conceptual knowledge enables learners to describe and define concepts. Traditional classroom teaching has been criticized for focusing on the facilitation of conceptual knowledge at the expense of other knowledge types. Mandl, Gruber, and Renkl (1994a) claim, for instance, that learners are typically facilitated and accustomed to acquiring conceptual knowledge, but they are not supported and accustomed to acquiring other knowledge types.

### Procedural knowledge

"*Procedural knowledge* contains actions or manipulations that are valid within a domain" (De Jong & Fergusson-Hessler, 1996, p. 107). Procedural knowledge enables learners to analyze and solve problems. With respect to epistemic activities, the construction of relations between conceptual and problem space is comparable to procedural knowledge. In some approaches, aspects of what has been also known as 'procedural knowledge' have rather been understood as a quality of knowledge, which describes whether knowledge was applicable or inert.

### Strategic knowledge

*Strategic knowledge* is knowledge about the sequence of solution activities (De Jong & Fergusson-Hessler, 1996). Strategic knowledge enables learners to identify separate steps and their order towards a problem's solution. Strategic knowledge may be implicit to a sequence of epistemic activities or emerge as a non-epistemic activity when collaborative learners coordinate group activities. 'Script' may be another term for strategic knowledge (Kollar, Fischer, & Hesse, 2006; Schank & Abelson, 1977). This notion of 'script' differs from what is meant by externally induced cooperation scripts as an instructional approach. Externally induced cooperation scripts are scaffolds to structure learners' interactions as an intended instructional support. In contrast, scripts as strategic knowledge describe cognitive structures, which may be called 'internally represented.' Kollar et al. (2006) link these script terms and refer to knowledge as distributed over an environment (cf. Salomon, 1993b). In this way, externally induced cooperation scripts are a kind of manifestation of strategic knowledge. This two-fold use of the term 'script' will be discussed further in section 4.3.

### 3.1.2 Qualities of knowledge

In addition to types of knowledge, De Jong and Fergusson-Hessler (1996) discuss several knowledge qualities, namely level of knowledge, structure of knowledge, automation of knowledge, modality of knowledge, and generality of knowledge. Some of these qualities have clearly been referred to as good vs. poor knowledge. This normative approach towards knowledge quality may be functional for educational psychologists and facilitators to set goals of pedagogical interventions. For instance, knowledge may be deep or superficial with regard to the level of knowledge. This evaluation of knowledge quality is based on differences between 'good' expert knowledge and 'poor' novice knowledge. Experts do not differ much from novices regarding superior memory or specific types of knowledge, but rather by superior knowledge qualities.

#### Level of knowledge

Knowledge is often described as *deep* or *surface-level*. A deep level of knowledge is characterized as the understanding of basic concepts, principles, or procedures and enables learners to take multiple perspectives about a problem (Snow, 1989). In contrast, surface-level knowledge is associated with reproduction and rote learning (Glaser, 1991). Surface and deep knowledge refer to the question of whether or not a learner recognizes surface features of a problem or has deep knowledge about problem features that are not apparent (Chi & Bassok, 1989; Dufresne, Gerace, Thibodeau Hardiman, & Mestre, 1992). In a first step experts recognize deep features of problems and identify the applicable principles, concepts, and procedures. Only as a second step, do experts concretely apply the procedures to solve a problem. Novices, in contrast, aim to identify surface features of a problem and compare them with a surface-level goal state. Often, novices will then immediately engage in concrete operations to reduce the distance between a problem's initial state and goal state (Dufresne et al., 1992).

### Structure of knowledge

Structure of knowledge has been argued to be the main difference between expert knowledge and novice knowledge (Larkin, McDermott, Simon, & Simon, 1980). Experts chunk information together into larger, more meaningful units to build a *hierarchic knowledge structure*, known as schemata or scripts (Chi, Feltovich, & Glaser, 1981; Dufresne et al., 1992; Schank & Abelson, 1977). Knowledge may be hierarchically structured in reference to the importance of individual knowledge components. A hierarchic knowledge structure is suited best for retention, for quick and efficient search processes, and for accomodating new knowledge (Boshuizen & Schmidt, 1992; Reif & Heller, 1982). This chunking of information has been well researched with recall tasks, where subjects are shown nonsensical problem-state configurations (e.g., when the pieces on a chess board are arranged in a random configuration). When it is not possible to chunk information, recall performance of experts and novices is equally poor (Dufresne et al., 1992; Gruber, Renkl, & Schneider, 1994).

### Automation of knowledge

Another difference between novices and experts is that novices use their knowledge by conscious, stepwise processes. Novices often need to make their knowledge explicit. Experts, in contrast, are claimed to make use of their knowledge in a continuous, fluid, and automatic process (De Jong & Fergusson-Hessler, 1996). Expert knowledge is therefore often referred to as tacit or implicit knowledge. It has been argued that this quality of expert knowledge may be acquired through informal learning, i.e. accumulating experience or mentoring, typically outside of educational institutions (Gelman & Greeno, 1989).

### Modality of knowledge

Based on Paivio's (1986) dual coding hypothesis, De Jong and Fergusson-Hessler (1996) suggest that knowledge can be stored in long-term memory as a set of propositions or images. The dual coding hypothesis argues that knowledge is more easily remembered when it is represented in the mind in multiple codes. Words and pictures may facilitate dual coding, but not necessarily implying that words foster internal representation as a set of propositions and that pictures foster internal representation as images. For instance, concrete words are represented in both codes. The word "table" typically activates a specific set of propositions as well as an image of a table.

### Generality of knowledge

Knowledge may be more or less transferable between tasks. For instance, heuristics may be more or less domain independent. There has been some research on facilitation of domain independent problem-solving strategies (e.g., Schoenfeld, 1985) but there is a growing number of studies which emphasize domain dependence of expert knowledge (cf. Gerstenmaier & Mandl, 2001; Mandl, Gruber, & Renkl, 1991).

## 3.2 Knowledge as Co-Construct and as Individual Acquisition

Knowledge is usually examined with respect to the individual. Hence, knowledge is often misconceived as "*furniture* of the [individual] mind" (Yale Corporation, 1828, p. 7). Knowledge can also be described as a more or less dynamic state of mind. Knowledge is characterized as a process and activity state in task performance, instead of an object of mind. Knowledge emerges in use only (De Jong & Fergusson-Hessler, 1996), which concludes when collaborative learners solve problems on grounds of shared knowledge or when individual learners perform tasks based on knowledge that they have acquired. In this respect, co-constructs of the group and individual knowledge acquisition, have been understood as two distinct concepts (Crook, 1995; Dillenbourg, 1999; Fischer, 2001; Hertz-Lazarowitz, Benveniste Kirkus, & Miller, 1992; Slavin, 1992). In collaborative knowledge construction, the analysis of both knowledge as co-constructed by the group as well as acquired by the individual has been suggested (cf. Means & Voss, 1996).

### 3.2.1 Knowledge as co-construct

Knowledge as a co-construct is knowledge that emerges in and results from collaborative knowledge construction (Fischer, 2001). In collaborative knowledge construction, learning has been regarded to as by-product of collaborative problem-solving processes (Dillenbourg, 1999; Gagné, 1987). Learners who construct knowledge together apply knowledge to complex problems; they produce a more or less adequate solution to a problem together. This group performance reflects the use of knowledge as co-construct. In this way, co-constructs of the group are subject to *effects with* the shared learning environment (Salomon & Perkins, 1998). Due to a specific collaborative learning environment, the group of learners may work together on a task more or less efficiently.

### 3.2.2 Individual acquisition of knowledge

Some socio-cognitive perspectives hold knowledge as embedded in social contexts. The individual mind is still considered as the agent of cognition, however, and individual knowledge acquisition is the predominant goal of any learning environment (Salomon, 1993a). Salomon and Perkins (1998) discuss individual learning outcomes as *effects of* a specific learning environment. This concludes that knowledge is acquired and possibly transferred to different situations by the individual. Learners need to transfer knowledge from problems worked on collaboratively to problems confronted with individually. The lack of spontaneous transfer has been well researched as a central problem of learning (cf. Mandl et al., 1994b).

## 3.3 Focused and Multi-Perspective Applicable Knowledge

Typically, learners are not facilitated to systematically apply knowledge from multiple perspectives in traditional classrooms, but to memorize isolated surface features of conceptual knowledge (Resnick, 1987). In this way, knowledge remains inert in traditional classroom teaching. Collaborative knowledge construction has been argued to facilitate qualities of knowledge that are typically disregarded in traditional classroom teaching. For instance, multiple perspectives and applicable knowledge have been argued to pose central benefits of collaborative knowledge construction (Renkl et al., 1996; Spiro et al., 1989). Still, *applicability* and *perspectivity* of knowledge have not been systematically researched. These two qualities of knowledge need to be discussed in more detail.

### 3.3.1 Applicability of knowledge

Applicability of knowledge, which has been regarded to as central quality of knowledge by some researchers (Bransford et al., 1989; Mandl et al., 1994b), can be regarded to as a question of knowledge structure (De Jong & Fergusson-Hessler, 1996). Structured expert knowledge is effectively applied. Compartmentalized novice knowledge, in contrast, is often inaccessible and cannot be put to use. Several studies more closely examined *how* knowledge is being applied. For instance, Fischer, Gräsel, Kittel, and Mandl (1996) examined coherence and accuracy of applicable knowledge of students solving complex cases. The starting point of this study was that students rarely build a coherent representation of a case. Complex problem cases contain much case information to which several theoretical concepts need to be applied (Gräsel & Mandl, 1993). Learners with compartmentalized knowledge, in contrast to structured knowledge, may not be able to build relations between case information and theoretical concepts.

### 3.3.2 Perspectivity of knowledge

Perspectivity of knowledge corresponds with the level of knowledge in De Jong and Fergusson-Hessler's (1996) matrix. Level of knowledge has often been used in a normative way and can include indistinct ideas of good (or deep) versus poor (or surface) knowledge. Perspectivity can be referred to as one more concrete quality aspect of knowledge. Regarding perspectivity of knowledge, two qualities can be distinguished: learners may consider one perspective and apply *focused knowledge*, or learners may consider multiple perspectives and apply *multi-perspective knowledge*. This does not conclude that multi-perspective knowledge is 'good,' and focused knowledge is 'bad' knowledge. Both knowledge constructs can be referred to by their distinct epistemic functions.

The few conceptual approaches regarding these qualities of knowledge mainly stem from medical education of diagnostics (cf. Kassirer, 1995). Therefore, examples from research of learning to diagnose will be referred to in the following paragraphs.

*Focused applicable knowledge*

Focused applicable knowledge is knowledge concerning the central aspects of a complex problem case and how to interpret them. Novices are rarely able to adequately relate their knowledge to problem cases. In medical diagnoses, for instance, so-called elementary findings (first basic findings like pale skin, fatigue etc.) are often interpreted in an inadequate theoretical reference frame or are not organized to form a coherent clinical picture (Fischer et al., 1996). Focused applicable knowledge is necessary to recognize the central aspects of a problem case and to apply the respective theoretical concepts to these aspects. Said differently, focused applicable knowledge directly addresses the core problems of a case.

### Multi-perspective applicable knowledge

Multi-perspective applicable knowledge is knowledge concerning the adequate explanation of case information which can refine the analysis of a problem case through the aid of theoretical concepts. Hereby, contradicting case information is also being considered for a complete analysis of a case. This form of applicable knowledge can thus also be used as premise regarding the extent to which learners are able to consider alternative explanations for a problem case (Vye et al., 1997). In medical diagnoses, for instance, various findings may support a range of clinical pictures. Novices, however, often generate final diagnoses based only on elementary findings, disregarding further case information, which may specify or qualify a first diagnosis. In this way, alternative diagnoses are being dismissed and final diagnoses reflect a vague clinical picture (Joseph & Patel, 1990). Novices seldom apply knowledge adopting multiple perspectives on a range of sub-problems (Vye et al., 1997). Multi-perspective applicable knowledge enables learners to consider various alternatives and to specify an analysis in specific directions.

## 3.4   Conclusions and Limitations: Dimensions of Applicable Knowledge as Outcomes of Collaborative Knowledge Construction

Knowledge often remains inert. Recent instructional approaches therefore have aimed to facilitate applicable knowledge (e.g., Dochy et al., 2003). Applicable knowledge has been considered as knowledge as co-construct, and as individually acquired knowledge (Fischer et al., 2000; Salomon & Perkins, 1998). Furthermore, qualitatively different levels of applicable knowledge can be pointed out as being focused or multi-perspective. Learners may apply knowledge to central aspects of a problem or consider various sub-problems. These forms of applicable knowledge have been regarded to as being particularly facilitated by collaborative knowledge construction.

With respect to applicable knowledge, a knowledge-in-use metaphor has been taken into consideration (De Jong & Fergusson-Hessler, 1996). Knowledge is not described as a definitive object of mind, but rather as an activity state of cognitive structures when individuals or groups perform tasks. The authors themselves of this approach qualify the selectivity of some types and qualities of knowledge. Some qualities of knowledge overlap or contain a range of sub-concepts (e.g., deep knowledge is characterized by multiple perspectives). De Jong and Fergusson-Hessler's (1996) matrix may therefore be regarded to as a non-deterministic scheme. This scheme allocates various forms of knowledge with the aim to systematize the many concepts of knowledge. Thus construct inflation can be reduced and researchers can be facilitated to share conceptions of what knowledge means. In this way, focused and multi-perspective applicable knowledge can be specified as knowledge variant in depth and structure.

Another limitation regarding the notion of knowledge-in-use is that it is to some extent self-referential. Similar to the circular idea that intelligence equals its measurement (Boring, 1923), knowledge-in-use could be defined by its measurement. This concept reflects some of the uncertainty of research about knowledge. Indeed, the individual knowledge types are associated with typical measurements, such as multiple choice tests for conceptual knowledge, or complex problem tasks for applicable knowledge.

# 4 Instructional Support for Collaborative Knowledge Construction: CSCL Scripts

A vast amount of energy and resources has been invested in the last decade to develop and establish CSCL in schools and universities (e.g., connecting schools to internet – "Schulen ans Netz"-initiative in Germany). Typically, computers have been adapted to traditional classroom practices. In this way, computers have not been systematically utilized to improve educational practices based on approaches and findings of educational psychology (Gräsel & Fischer, 2000). Regarding the use of computers for the improvement of education *CSCL approaches based on media* are needed on one hand. These approaches consider the specifics of CSCL environments in order to realize instructional support. On the other hand, *instructional approaches with an educational psychological background* are needed. These approaches utilize knowledge that has been acquired through decades of research on collaborative knowledge construction.

Some instructional approaches aim to establish beneficial conditions for learning together such as *training the individual learner to cooperate* in order to facilitate collaborative knowledge construction (King, 1994; Rummel & Spada, 2005; Webb & Farivar, 1994). However, training of collaborative skills has been argued to be costly and impractical in CSCL environments (Weinberger, Reiserer, Ertl, Fischer, & Mandl, 2005). Some of these training programs, for instance, take more time than the actual collaboration of learners (e.g., Hytecker, Dansereau, & Rocklin, 1988). Furthermore, online learners may not be able to participate in FTF training programs.

There are alternatives to establishing beneficial prior conditions. Process characteristics that influence collaborative knowledge construction can be identified in order to conceptualize direct instructional support for these processes. Based on socio-cognitive perspectives, specific social and cognitive processes can be identified as essential characteristics of collaborative knowledge construction. There are a number of instructional approaches aiming towards specific processes of collaborative knowledge construction (e.g., King, 1999). However, there is little systematic research on how CSCL environments may be designed based on approaches of educational psychology to facilitate knowledge construction.

Instructional approaches to facilitate social and cognitive processes in CSCL environments can be based on acknowledged approaches of educational psychology. In this way, instructional support can be conceptualized, which does not only aim to compensate differences between CMC and FTF discussions, but to facilitate specific social and cognitive processes. The medium may not be deficient per se in comparison to FTF discourse. Instead, text-based CMC may afford and constrain activities in ways that can be detrimental or beneficial to collaborative knowledge construction. The medium can therefore be understood as a natural resource for collaborative knowledge construction. Instructional support may aim to exploit this resource and simultaneously sublate the limitations of CSCL.

In this chapter, media-based instructional approaches specific to CSCL will be discussed first. Second, functions of CSCL environments, which facilitate collaborative knowledge construction, will be introduced. Third, cooperation script approaches will be introduced in detail as a theoretical background rooted in educational psychology for facilitating collaborative knowledge construction in CSCL environments. Next, implementation of cooperation scripts into CSCL environments will be discussed. Subsequently, social and epistemic cooperation scripts will be introduced as instructional support that can be realized in CSCL environments. Finally, problematic aspects of the cooperation script approach will be discussed.

## 4.1 Facilitating CSCL by Media Choice and Interface Design

Approaches to facilitate CSCL are typically based on the possible influence of the computer interface as medium for collaborative knowledge construction. These approaches of CSCL research may help to regard media-specific aspects of facilitation. CSCL research considers specific functions of CSCL environments and provides a number of techniques concerning the realization of support in CSCL environments. One CSCL approach is to examine how the use of various media may influence collaborative knowledge construction (Schweizer, Paechter, & Weidenmann, 2001). Another approach is to modify media to suit specific purposes (Hesse et al., 1997). Collaborative knowledge construction in CSCL environments may therefore be facilitated by *choosing the most adequate medium* for the specific learning scenario or by *adapting media interfaces* to serve specific purposes, such as to foster specific processes of collaborative knowledge construction (Weinberger & Mandl, 2003).

### 4.1.1 Media choice approaches

The adequate *media choice* may appear to be a simple and obvious approach to facilitate collaborative knowledge construction, because any media may include advantages and disadvantages related to different scenarios of collaborative knowledge construction. This *task-media-fit-approach* has been mainly researched in CSCW (Computer-Supported Collaborative Work) rather than CSCL (McGrath & Hollingshead, 1994). Results of CSCW research may also provide valuable input to CSCL that is based on problem-solving. The task-media-fit approach involves the notion that, for instance, text-based CMC may be more appropriate for some tasks than FTF-communication. The individual capacity (bandwidth) to transmit more or less information through these media is matched with a number of tasks that require different degrees of information (McGrath & Hollingshead, 1993, 1994). An idea-generating task, for instance, does not require as much interaction between the discussants as tasks that involve the negotiation of conflicts, e.g., the task to solve complex problems. Therefore, idea generating tasks are appropriate for text-based CMC whereas other tasks require more bandwidth as provided in video conferencing or FTF-communication. The task-media-fit-approach therefore suggests that media choice should be *rational* to facilitate collaborative knowledge construction. A rational media choice means that specific characteristics ascribed to the individual media make the media more or less appropriate for specific communicative scenarios. For instance, email has been judged as appropriate for information exchange, but in order to get to know each other, FTF-communication is usually considered more appropriate (Rice, 1993). However, text-based CMC, e.g., in web-based discussion boards, has also been considered to foster specific social and cognitive processes. Generally, text-based CMC has been argued to facilitate more reflective discourse and thus 'deep' processing of knowledge (e.g., Marttunen & Laurinen, 2001). Actual media choice in real world settings may not always be rational, however, and may not always be based on an ideal fit between medium and collaborative knowledge construction scenario (cf. Döring, 1997a).

Media may be chosen *normatively* based on what users know and appreciate best. It has been shown, for instance, that the appreciation of email in organizational contexts is related to the experience of the individual in handling email and also depends on how colleagues and superiors appraise email (Schmitz & Fulk, 1991). This includes the notion that the preferred medium is not necessarily the most costly, high-bandwidth medium. Anderson and colleagues describe, for instance, that video conferencing may be considered less useful compared with non-interactive video resources (Anderson et al., 2000). Normative media choice may explain, why ready to use video conferencing technology has existed for more than 20 years, but 'videophones' are still sparsely used. In contrast, the text-based SMS (Short Message System) has quickly become a wide spread modern mobile communication option in spite of the fact that technically more advanced forms of communication exist.

Media choice may also be *interactive* and depend on how many and to what extent possible communicants use a specific medium. In this respect, a critical mass of communicants enhances the use of a specific medium (Markus, 1987). Some studies show that communities can be supported best by using modest, common, and easily accessible equipment rather than hi-tech, highly specialized communication tools (Carletta, Anderson, & McEwan, 2000). Therefore, computer-based approaches to facilitate collaborative knowledge construction should consider the actual media context of the individual learner.

### 4.1.2    Interface design approaches

CSCL may be facilitated by *interface design*. This approach argues that no medium was genuinely designed for collaborative knowledge construction and thus, the design of the medium interface could be modified and improved for specific CSCL scenarios (Hesse et al., 1997; Mandl & Fischer, in press; Roschelle & Pea, 1999). Media can therefore be adapted to foster collaborative knowledge construction by technically implementing support into the CSCL environment. The development and experimental research of interface design to support collaborative knowledge construction has many practical implications. Typically, CSCL research strives to design interfaces which reduce the deficits of CMC as compared to FTF-communication. Many interfaces have been designed, for instance, to reduce the coordination disadvantages of CMC (cf. Hesse et al., 1997). Interfaces may also be designed to foster specific interactions found to be beneficial for collaborative knowledge construction in educational psychology (e.g., Hron, Hesse, Reinhard, & Picard, 1997). The rationale is that a specific interface design may endorse or even substitute extensive training and feedback by co-present moderators and warrant a standardized quality of collaborative knowledge construction (cf. Collins, Brown, & Newmann, 1989). The interface may afford and constrain specific activities of collaborative knowledge construction. In this way, specific functions of media may be facilitated or reduced in order to facilitate collaborative knowledge construction.

## 4.2 Functions of Media for Collaborative Knowledge Construction

Media can acquire several functions which facilitate collaborative knowledge construction (Mandl & Fischer, in press; Roschelle & Pea, 1999). These functions may be inherent to some media but may also be enhanced by specific interface design. These functions have been categorized as shared representation function, community-building function, and structuring function.

### 4.2.1 Shared representation function

An interface may provide discussants with shared representations of the subject matter through different codes (text, graphic, etc.). Representations may guide collaborative knowledge construction by emphasizing specific aspects of a subject matter (Fischer et al., 2002). The salience of specific aspects in representations would increase the chance that these aspects would enter the discourse. Representations may also facilitate collaborative knowledge construction by providing a common ground of the discussants in accordance with the *physical co-presence heuristics* (Clark & Marshall, 1981). Shared representations are meant to function as a common reference point and complement discourse by providing information that does not need to be interpreted, but can be immediately used by the discussants (Mandl & Fischer, in press). In this respect, shared representations could reduce ambiguous communication. For instance, graphical representations may define subject matters in more definite, complete ways than was possible in discourse without graphical representations (Schnotz, Boeckheler, & Grzondziel, 1997). However, as Jucks, Bromme, and Runde (2003) show, representations could also increase the *illusion of evidence* in expert-novice communication. When explaining a subject matter, representations falsely suggest critical information to be obvious. Explainers overestimate the explanatory power of the representations. For instance, physicians may often overestimate how much their patients recognize in radiographs and thus neglect to fully explain the radiograph. In asynchronous CMC this illusion of evidence may become more problematic, because discussants cannot immediately give feedback of incomprehension (Bromme & Jucks, 2001).

This inconsistency (representations facilitating common ground vs. representations creating illusion of evidence) may be explained by the fact that the beneficial effects of shared representations are highly dependent on the degree of prior knowledge of all communicants (Fischer, 1998; O'Donnell & Dansereau, 2000).

Shared active representations may also model group processes and the subject matter in a more interactive way. For instance, *mapping techniques* are based on the idea of representing individual concepts on single cards and graphically linking these concepts with specified relations on a map. Online mapping techniques have been successfully applied in CSCL environments (Fischer et al., 2002). The rationale of these more interactive forms of shared active representation tools is that users may record important processes and results of collaborative knowledge construction. This permanent record may in turn facilitate collaborative knowledge construction. Therefore, discussants may be less likely to fall victim to an illusion of evidence when they need to construct a shared representation together. Typically, these mapping techniques are not designed for specific tasks and are content-unspecific. In Fischer et al.'s (2002) study, content-specific vs. content-unspecific mapping techniques in CMC based on video conferencing were varied. The results show that in contrast to content-unspecific mapping techniques, content-specific mapping techniques foster the construction of conceptual space and the construction of relations between conceptual and problem space as well as externalization, elicitation, and conflict-oriented consensus building. Thus it can be said that mapping techniques, which are designed for specific purposes, can improve CSCL.

## 4.2.2    Community-building function

CSCL environments may support the social coherence of communities by providing defined virtual spaces as MUDs, for instance, do (*MUD = Multi User Dungeon*, Dillenbourg, 2002; Mandl & Fischer, in press). MUDs consist of textually represented spaces, objects, and characters or 'avatars.' Community-building in CSILE, for example, is based on the principle that the individual members of the community contribute to a specific subject matter on which other members of the community may further build (Scardamalia & Bereiter, 1996). In this way, the community-building function of interfaces helps to allocate knowledge resources, to build groups of interest and to continuously generate better answers to complex problems.

One aspect of community-building is therefore *knowledge mining*. This means that the community-building function may help users to discover the knowledge of the entire community on a particular subject matter and recommend specific resources and experts within the community (Roschelle & Pea, 1999). The separation of the CSCL environment into specific, purpose-built virtual spaces, which may be accessible only by community members, aims to improve the knowledge search within a community (Weinberger & Lerche, 2001). For instance, "online-cafés" are supposed to provide space for informal conversations, "virtual information centers" inform new community members how to use the environment, "virtual libraries" represent the collected archive of the community, etc. In this way, the community-building function of interfaces aims to support the accumulation of structured knowledge. Thus, the value of the individual contributions of the community members is intended to be augmented. Individual contributions are made accessible to a community and framed in a growing shared knowledge structure (Roschelle & Pea, 1999).

The community-building function can be regarded to as central to CSCL environments in real world practice. Mandl and colleagues have conceptualized several design principles for computer supported learning communities (Fischer & Mandl, 2002; Reinmann-Rothmeier, 2003; Reinmann-Rothmeier & Mandl, 2002; Reinmann-Rothmeier & Mandl, 2001; Weinberger & Lerche, 2001; Weinberger & Mandl, 2003; Winkler & Mandl, 2002).

These design principles include, for instance, embedding FTF phases into online seminars, structuring online processes, or providing task-specific virtual spaces.

### 4.2.3      Socio-cognitive structuring function

Interfaces may be designed to structure discourse aiming to induce successful patterns of collaborative knowledge construction. Successful interaction patterns may involve specific social and cognitive processes. For instance, learners may be guided through a CSCL environment along prescribed paths. Thus, more specific, individual activities can be suggested by interface design. Typically, CSCL research aims to compensate differences between CMC and FTF collaboration by facilitating coordination in CSCL environments.

Baker and Lund (1997) pre-structured interactions of learners in a detailed manner by providing buttons for specific speech acts in a text-based CMC interface of a CSCL environment. The buttons are labeled with speech acts, such as "I propose to ...," "Ok," "Wait!" etc., that could be pasted into the interface and eventually completed by the user. Learners were expected to use those buttons to reduce typing demands. Some speech acts would also improve socio-cognitive knowledge construction processes and grounding. Hron et al. (1997) sequenced the interaction of learners by alternately prompting learners to propose correction of the learning partner, explain the correction, and obtain agreement from the learning partner. Only when both partners reached agreement was the interface accessible to realize the correction.

In comparison to discourse in text-based CMC, without structuring induced by the interface, the studies show that structuring interfaces can improve processes of collaborative knowledge construction and encourage discussants to disagree and explore alternative viewpoints (e.g., Pfister & Mühlpfordt, 2002). Furthermore, it has been found that structuring facilitates epistemic activities (e.g., Baker & Lund, 1997; Hron, Hesse, Cress, &

Giovis, 2000). In this respect, text-based CMC may be appropriate for directly modifying discourse by sequencing and timing content or interaction. Structure may also be induced with text-based CMC by assigning specific activities or roles to individual group members. The focal point of realizing structuring functions in CSCL is to make up for coordination disadvantages in CMC. Structuring functions may also aim at specific processes of collaborative knowledge construction based on instructional approaches rooted in educational psychology. In this way, collaborative knowledge construction in CSCL might be improved beyond what could be achieved even by well-coordinated collaborative learners.

## 4.3　　　Script Approaches

Educational psychology may provide approaches to systematically enhance the socio-cognitive structuring function of CSCL environments. A number of script approaches have been developed based on empirical findings and socio-cognitive theories to directly facilitate specific *processes* of collaborative knowledge construction (see Huber, 1999).

Instructional script approaches are based on prescriptions for learners, induced by educational facilitators, to facilitate specific knowledge construction activities. The script term, however, has also been used to describe cognitive structures guiding individuals through specific processes (e.g., *restaurant script*, Schank & Abelson, 1977). That means, that individuals possess knowledge about what specific activities in what sequence need to be applied in a specific context. In a typical restaurant, for instance, individuals know that they first need to order a menu, subsequently wait to be served and eat, and finally pay the bill. Kollar et al. (2006) link Schank and Abelson's (1977) idea of scripts as cognitive structures and instructional script approaches (O'Donnell & Dansereau, 1992). Collaborative knowledge construction may be always more or less guided by scripts, but not all scripts may be externally induced by facilitators. Learners may interact on grounds of cooperation scripts that are represented in their minds instead.

For instance, learners may decide to compete against each other to contribute the best solution to a problem. These scripts are not explicitly introduced by facilitators, but are already represented in cognitive structures and activated by cultural practice. These internally represented scripts may structure the social and cognitive processes of collaborative knowledge construction. For instance, learners may work on a set of subtopics of a complex problem. Some of these internally represented scripts have been argued to be suboptimal for collaborative knowledge construction (Person & Graesser, 1999). Cooperation scripts, which are induced by educational facilitators, however, have generally shown to be a promising approach to foster collaborative knowledge construction (e.g., Rosenshine & Meister, 1994).

The underlying principles of script approaches are to *specify*, *sequence*, and *assign* activities to collaborative learners (cf. Dansereau et al., 1979). *Specifying activities* should help learners to produce activities which are beneficial to collaborative knowledge construction and to avoid activities which may be detrimental. Typically, a teacher specifies activities, which are believed to facilitate knowledge construction, prior to a collaborative phase of learners. For instance, teachers introduce students to the collaborative learning strategy of question asking. Subsequently, learners are expected to engage in the specified activities in the collaborative phase. Furthermore, *sequencing activities* aims to support productive interactions. The specified activities may be beneficial for collaborative knowledge construction only when they are applied at specified times. Sequencing should warrant that the students engage in the specified activities at specific times. In this way, the specified activities may be organized to build sensible discourse structures. For instance, after question asking, the sequence of a script may suggest to answer questions as a next step. Therefore, sequencing may support learners to better relate to each other and support transactive discourse. *Assigning activities* aims to warrant that the specified activities are carried out by all learners. This typically includes that learners are expected not only to engage in one specific activity, but to take turns in assuming responsibility for various specified activities. For instance, one learner may be assigned the activity to ask questions regarding one specific problem case and another learner may be expected to answer those questions. Subsequently, these learners may switch their roles to work on a following problem case.

Scripts may be realized in several ways. Typically, scripts are introduced and monitored by teachers. Therefore, script learning includes phases of traditional classroom teaching wherein learners are taught to apply script structures. This may let down the original idea of collaborative knowledge construction as an alternative to traditional classroom teaching to some extent. Therefore, some script approaches aim to structure specific activities differently. For instance, collaborative learners can be provided with artifacts (= man-made objects). Artifacts may be designed to carry information about specific activities. When collaborative learners use these artifacts together, they may automatically engage in a sequence of activities.

In this section, the original scripted cooperation approach will first be discussed (O'Donnell & Dansereau, 1992). The primary question is, which activities scripts should specify? In reference to important processes of collaborative knowledge construction, scripts may aim to facilitate social and cognitive processes. Subsequently, another script approach, which is guided peer questioning, will be introduced (King, 1999). Guided peer questioning makes use of artifacts to suggest specific activities to learners.

### 4.3.1    Scripted cooperation

The first instructional approach to coin the 'script' term is the *scripted cooperation* approach (Dansereau, 1988), which aims to facilitate text comprehension. The main principle of scripted cooperation is to provide instructions for learners to engage in specific activities in collaboration. In this way, activities beneficial to collaborative knowledge construction should be facilitated and activities that are detrimental to collaborative knowledge construction may be limited. First of all, a text will be segmented. Then, the MURDER-script specifies several activities for two learners (cf. O'Donnell & Dansereau, 1992):

- <u>M</u>ood – the learners relax and concentrate on the task
- <u>U</u>nderstand – both partners read the first section of the text
- <u>R</u>ecall – learner A reiterates the text section without looking at the text
- <u>D</u>etect – partner B provides feedback without looking at the text
- <u>E</u>laborate – both learners elaborate on the information
- <u>R</u>eview – both partners look through the learning material once again

The learning partners are supposed to engage in these activities for each text segment, switching roles regarding recall and detection for each segment, until they have completed the text.

Several variants of the prototypical scripted cooperation approach have been developed and examined as the prototypical script confounds several dimensions of important co-construction activities (cf. Huber, 1999). The activities suggested by the MURDER script aim, for instance, at affective, elaborative, as well as meta-cognitive activities. Early attempts to disentangle the confounding of several dimensions of collaborative knowledge construction have been made (Larson et al., 1985; see also O'Donnell et al., 1987). Larson et al. (1985) compared effects on the amount and accuracy of knowledge as co-construct and individually acquired knowledge of an elaborative and a meta-cognitive cooperation script. This comparison showed diverging effects on knowledge as co-construct and individually acquired knowledge. The meta-cognitive cooperation script of this study produced a positive effect on knowledge as co-construct, but was detrimental for individually acquired knowledge. The elaborative cooperation script, in contrast, only facilitated individually acquired knowledge, but impeded knowledge as co-construct. In later works, O'Donnell and Dansereau (1992) differentiated several more dimensions of collaborative knowledge construction activities, namely cognitive, affective, meta-cognitive, and social dimensions. These dimensions may be confounded in individual scripts. In several contributions, an equally multi-dimensional approach to facilitate learning has been suggested. Many researchers particularly suggest facilitating social and cognitive dimensions of collaborative knowledge construction (Mandl et al., 1993). In a recent contribution, O'Donnell (1999) made "two key assumptions" with respect to scripted cooperation for collaborative knowledge construction:

> First, the use of scripted cooperation will prompt the use of *cognitive*
> *processes* by participants that might otherwise not occur. For example
> during scripted cooperation, students explicitly engage in error detection
> when they might not routinely do so. Second, the use of scripted coopera-
> tion can limit the occurrence of negative *social processes* that may im-
> pede group functioning and achievement. For example, students often as-
> sume a single role and maintain that role throughout a study period. (p.
> 180, italics added by the author)

Prototypical cooperation scripts aim to structure both social and cognitive processes. The facilitation of social and cognitive processes should
facilitate collaborative knowledge construction, because learners often discuss at a superficial level and digress or argue about isolated and naïve concepts (e.g., Hogan et al., 2000), and because learners' spontaneous cooperation strategies often prove to be sub-optimal (e.g., Webb, 1989). Little
is known, however, regarding cooperation scripts that aim either at social or
at cognitive processes of collaborative knowledge construction. There are
only a few studies that have analyzed scripts which explicitly aim at these
specific dimensions (Dufresne et al., 1992; King, 1992; Palincsar & Herrenkohl, 1999). The results of these studies may convey an impression of
how the distinct social and cognitive processes can be facilitated. These
studies will be illustrated in more detail in the following paragraphs.

*Scripts aiming to facilitate social processes.* Social cooperation
scripts typically support specific roles in order to facilitate those social
processes, which are meant to be related to knowledge construction but
rarely emerge spontaneously in collaborative knowledge construction. Elicitation, for instance, has been shown to be a discourse activity that peer
learners do not engage in sufficiently for learning purposes (King, 1990).
Therefore, King (1992) has aimed to facilitate this social mode of co-
construction with the help of social scripts. She found that when elicitation
is facilitated successfully, individual knowledge acquisition is also fostered
(cf. section 4.3.2; King, 1999). Further social processes may be found sub-
optimal and have been facilitated with social scripts. Conflict-oriented consensus building, for instance has been linked to knowledge acquisition (e.g.,

Chan et al., 1997), but has also been found to rarely emerge spontaneously in peers' collaborative knowledge construction (Tudge, 1992).

*Scripts aiming to facilitate cognitive processes.* Scripts aiming to facilitate cognitive processes may be more or less domain- and task-specific (Dufresne et al., 1992; King, 1999; Palincsar & Herrenkohl, 1999). Dufresne et al. (1992), for instance, rank questions with the help of a computer-supported learning environment according to a hierarchy, whereby the questions become increasingly problem specific. First of all, the learners are asked to select and define a principle that could be applied to solve the problem under consideration. Subsequently, the questions guide the learners to apply the theoretical principle to the problem. In this way, they aim to facilitate learners to mimic an expertlike, hierarchical problem-solving approach. Learners could be successfully supported with this script, which aims at cognitive processes to produce more expertlike performance than control groups. Dufresne et al.'s (1992) interpretation of this result concluded that the script aiming at cognitive processes helps to focus learners on expert task strategies and that these scripts may reduce cognitive load. In collaborative knowledge construction, scripts may aim to keep learners on-task, highlight solution procedures, and prompt learners to produce these specific problem-solving activities. There are indications, however, that expertlike strategies to solve problems in a top-down manner may not be a functional way for novices to acquire knowledge (cf. Gräsel, 1997).

In another study, "cognitive tools" and "intellectual roles" have been applied in fourth grade classrooms (Herrenkohl & Guerra, 1998). The cognitive tools provided task strategies including predicting and theorizing, summarizing results, and relating predictions and theories to results. These task strategies were introduced to each student as the framework for discussion prior to collaborative class activities. The intellectual roles did not provide additional strategies, but assigned the cognitive tools to individual learners, meaning that each student was made responsible for either predicting and theorizing, summarizing, or relating theory-based predictions and results. Classroom discussions of learners who were provided with intellectual roles in addition to cognitive tools engaged in more transactive discussion and better epistemic activities, which led Palincsar and Herrenkohl (1999) to argue

> [...] that providing a set of tools to guide students in constructing scientific explanations is not sufficient to ensure high levels of engagement and collaboration. To deeply engage students with the cognitive content and with other participants in the classroom, they need to be given roles with concomitant rights and responsibilities. (p. 169)

Overall, the results of the studies indicate that scripts aiming to facilitate cognitive processes may need to be carefully designed. Expertlike problem-solving strategies may not always provoke knowledge construction. Furthermore, scripts aiming to facilitate cognitive processes may need to be endorsed by more social scripts components like assigning "intellectual roles."

### 4.3.2 Guided peer questioning

In addition to the original scripted cooperation, several other instructional approaches have been subsumed as script approach (cf. Derry, 1999; Renkl & Mandl, 1995). In contrast to the original scripted cooperation approach, some script approaches use artifacts in addition to training to further facilitate specific interactions of collaborative learners (e.g., Clark et al., 2003; King, 1999; Weinberger et al., 2005). King's (1999) artifact-based script approach, guided peer questioning, utilizes hand-held prompt cards to induce specific learner activities. This approach will be introduced in the following paragraphs.

King found that students often have difficulties to spontaneously ask task-related, thought-provoking questions (King, 1989b). As a solution to these student-problems, King developed guided peer questioning (King, 1992). King (1999) argues that "different types of interaction facilitate different kinds of learning" (p. 88). She suggests that higher level of learning also require higher levels of interactions between learners (King, 1989b). According to King, task-related questions in particular, are indicators of higher level interactions. As a consequence, with a number of subsequent studies, King aimed to teach students various strategies to ask task-related questions and developed the *guided peer questioning* approach (cf. King, 1999).

King (1999) discusses three different discourse patterns for which she provides specific *prompts*: complex knowledge construction, problem-solving, and peer tutoring. These prompts are unfinished sentences, typically question stems that learners are expected to respond to and complete. With respect to *complex knowledge construction*, learners are firstly trained to give elaborated answers. Then, learners select a few question prompts from a larger list and generate several content-specific questions by 'filling in the blanks.' These prompts are, for instance, "What does ... mean?", "Explain why ...", "What would happen if ...?" and so on. Next, students discuss a subject matter asking their questions and by giving elaborated answers. With respect to *problem-solving*, students are provided with specific question starters on hand-held prompt cards, e.g., "What is the problem?", "What do we know about the problem so far?", "What is our goal now?", and so on. Two or three learners should then reciprocally construct questions with the help of the prompt cards and provide answers in small group discourse. The prompt cards are supposed to be flexibly used by students to mutually structure discourse. For *peer tutoring*, questions are provided for a tutor only in a specific sequence to assess and consolidate prior knowledge, construct new knowledge, and to monitor cognitive processes.

Results of empirical studies show that guided peer questioning is more effective regarding individual knowledge acquisition than discourse without instructional support (e.g., King, 1990). King's (1999) findings regarding processes of collaborative knowledge construction are mainly anecdotal. King (1999) describes that guided peer questioning prompted high level interaction, which includes activities such as asking thought-provoking questions, integrating new knowledge with prior knowledge, and examining alternative perspectives.

## 4.4 Blending Scripted Cooperation and Guided Peer Questioning: Prompt-Based Scripts

Both instructional approaches, scripted cooperation and guided peer questioning, may be combined to build the fundament for *prompt-based scripts*. The scripts in scripted cooperation have often been compared to theater scripts. Participants are asked to play roles in a specific order. However, the scripts in scripted cooperation do not really contain any text, that is they do not specify the roles during collaboration. In guided peer questioning some text is provided in form of question starters or prompts, but no script or description of roles is provided that puts these prompts in context. Prompt-based scripts combine both scripts and prompts. First of all, learners are briefly introduced to specific roles, the prompts, and to a scripted interaction structure. Then they individually work through a theoretical text. In a following collaborative knowledge construction phase, the learners are asked to apply theoretical concepts to several case problems together. The goal is to come to a joint solution of the individual cases. In this collaborative phase, learners are supposed to act out the individual roles in a specified sequence with the help of text-based prompts. The individual prompted roles and activities are complementary and are taken over reciprocally in a way, that all learners act out the various roles at specified times. In this way, roles and activities, as they are suggested in scripted cooperation, are supported by prompts, so that learners should not need to study the suggested individual roles and activities intensively, but rather (inter-)act according to what the prompts suggest. Therefore, prompts may help learners to fill out the specific roles and activities.

The rationale behind prompt-based scripts is that learners are guided through a series of activities without having to waste time and effort on preparing and thinking about specific roles, because the roles of the scripted cooperation approach (O'Donnell & Dansereau, 1992) are supported by prompts as in guided peer questioning (King, 1999). As Derry (1999) described King's approach, there is a "pleasing simplicity to this approach because it does not overload students' working memories with multiple goals and forms of training" (p. 205). Prompt-based scripts aim exactly at this kind of simplicity.

Some recent studies have examined processes and results of collaborative knowledge construction supported with prompt-based scripts. Interestingly enough, these studies have induced the prompt-based scripts in different ways. Typically, the prompts are realized with artifacts. For instance, prompts are written on cue cards that learners are confronted with. Ge and Land (2002) compared the *effect of question prompts on individual and collaborative learning* in FTF scenarios. Ge found that question prompts are especially effective in peer collaboration, but that individual learners can also be supported with question prompts with respect to problem representation, solutions, justifications, and monitoring and evaluation of learning processes.

Some studies have additionally provided training to use the prompts. Hron et al. (2000) compared *implicit structuring*, which induced group discussion on the subject matter by working on key questions in a preceding learning phase, and *explicit structuring*, which provided additional rules for discussion, namely to contribute equally and to engage in conflict-oriented discourse. These forms of structuring were provided by a facilitator before the collaborative, text-based computer-mediated setting. The results of this study show that both forms of structuring support epistemic activities and more transactive discourse.

Finally, a 'prompter' role may be assigned to one of the learners who is then responsible for inducing the prompts into discourse. Coleman (1998) investigated explanation prompts that were made available to learners on cards in FTF collaborative learning scenarios. These cards were also supposed to support roles. These roles were called *prompter* for the person who selects the prompt cards and places them in front of the learning partners, *reader and writer* for the person who would read the question and document the group work, and *explainer* for the person who generated explanations from the questions. The results of this study indicate that the prompts facilitated the construction of more advanced and correct explanations.

## 4.5     Implementing Prompt-Based Cooperation Scripts in CSCL Environments

Typical script approaches suitable for FTF collaborative knowledge construction may not simply be transferable to, but may rather be impractical for CSCL environments. Scripts for FTF collaboration carry certain costs of implementation, e.g., the costs of a moderator or facilitator who models problem-solving, monitors compliance with a script, or trains students prior to the actual peer interaction. In order to reduce these costs, prompt-based cooperation scripts are mainly realized by artifacts. These artifacts can be components of CSCL environments and aim to enhance the socio-cognitive structuring function. In this way, prompt-based cooperation scripts may be a feasible instructional approach for CSCL. Text-based CMC offers the possibility to structure the learners' discourse and can guide users through a certain series of activities by design of the interface (Baker & Lund, 1997; K. S. King, 1998; Nussbaum, Hartley, Sinatra, Reynolds, & Bendixen, 2002; Scardamalia & Bereiter, 1996). This means that the prompt-based cooperation script approach can be linked with CSCL approaches and realized with socio-cognitive structuring functions of specifically designed interfaces.

Cooperation scripts can be implemented into CSCL environments, for example, with the help of an interface providing a discourse structure and prompts that are inserted into text-windows of web-based discussion boards beforehand (Hron et al., 1997). For instance, Nussbaum et al. (2002) provided learners with a number of prompts called *note starters*, e.g., "My theory is ...." or "I need to understand," which students could choose when starting to write a message in text-based computer-mediated learning environments. These note starters are implemented into the text window, which discussants use to formulate messages in online debate. The findings of this study show that note starters could encourage students to disagree and explore alternative viewpoints in comparison to discourse without structure induced by interface design in text-based computer-mediated learning. Thus, it can be said, that prompts can have a positive effect on conflict-oriented consensus building in text-based CMC (Nussbaum et al., 2002). Furthermore, CSCL scripts seem to have a general positive effect on epistemic activities (Baker & Lund, 1997). Learners provided with CSCL scripts generally engage more in epistemic rather than non-epistemic activities. Little is known, however, about how scripts that aim at specific cognitive processes can be applied in CSCL and to what extent specific cognitive goals can be achieved with these CSCL scripts.

With respect to facilitating social and cognitive processes, most CSCL research is restricted to whether or not learners need to coordinate discourse moves and grounding (Baker & Lund, 1997; Hron et al., 1997). These studies show that coordination can be facilitated with CSCL scripts that are based on prompts and on structuring functions of CSCL environments. So far, there has been no systematic research with respect to CSCL scripts that are specifically oriented towards social and cognitive processes.

CSCL environments may be regarded to as a "natural" environment for the scripts approach. Prompt-based cooperation scripts can be realized in web-based discussion boards, for instance, rather than with hand-held prompt cards. Prompts may directly pre-structure the text messages of learners. Additional demands to implement the coordination of the prompts, e.g., with the role of a student prompter as Coleman (1998) suggests, may become redundant. Obviously, the facilitation of important processes of the collaborative construction of knowledge by artifacts has some advantages over instructional support that is realized by teachers (Weinberger & Mandl, 2001). The first advantage of this CSCL script approach is that the *quality of the instructional support is warranted* (cf. Collins et al., 1989). Individual teachers and students may interact in more or less beneficial ways. In contrast, CSCL environments can be continuously developed to better correspond with students' needs. Another advantage of applying CSCL scripts is that *students may apply scripts flexibly*. Typical script approaches determine student activities regardless of particular needs and individual learning prerequisites. Usually, moderators are monitoring that script prescriptions are accomplished by students. In CSCL environments students may decide more freely to what extent externally induced scripts are being followed or not. Lastly, *costs of instructional support may be reduced*. Other forms of instructional support suggest extensive training prior to actual collaborative learning as well as adaptive feedback of co-present experts. Costs for these forms of instructional support often pose a substantial barrier for successfully realizing collaborative knowledge construction.

## 4.6 Conclusions and Limitations: CSCL Scripts

In this chapter, it has been argued that discourse without structure induced by facilitators or interface design based on script approaches rarely produces interactions that are beneficial to knowledge construction. As a consequence, instructional support can be suggested that aims directly at the social and cognitive processes of collaborative knowledge construction. Cooperation scripts are instructional support, which is oriented towards processes and which specifies, sequences, and assigns activities and roles to collaborative learners. Cooperation scripts may be based on prompts as in guided peer questioning that support the individual activities suggested by a script. These prompts can support socio-cognitive structuring in CSCL environments. Thus far, support of social and cognitive processes has been confounded. Cooperation scripts may aim to facilitate what has been regarded to as important processes for collaborative knowledge construction. Therefore the development and implementation of an epistemic and a social prompt-based cooperation script into a CSCL environment can be suggested.

*Social cooperation scripts* aim to facilitate the specific social modes of co-construction (Fischer et al., 2002). Social scripts provide learners with roles and encourage them to perform particular interactions at specified times. These particular interactions may be related to specific social modes of co-construction. Thus, social scripts can aim to foster transactive discourse by supporting certain complementary roles such as constructive critic and case analyst. Social scripts can be based on a kind of schedule, in which any of the specified role activities can be allocated. An everyday example of such a procedure is the scientific peer review process. After a paper has been submitted, peers take over the critic role and point out deficits of a proposal. The scientific peer review process may not need additional instructional support, but can rather be understood as a social cooperation script, which is already represented in individual minds as well as in cultural practice sensu Schank and Abelson (1977). Novice learners, however, may need externally induced scripts, because they typically do not possess beneficial social cooperation scripts themselves.

*Epistemic cooperation scripts* aim to facilitate specific epistemic activities (Fischer et al., 2002). Epistemic scripts can aim to facilitate expertlike problem-solving behavior by engaging learners in epistemic activities that experts typically carry out (Dufresne et al., 1992). In this way, epistemic scripts are not supposed to add learning material, but can rather be understood as a kind of task strategy. Epistemic scripts should, for instance, help learners to consider all relevant aspects in an adequate order. An everyday example of such an epistemic script is a checklist. Checklists typically induce the task strategy to consider several domain- and task-specific aspects in a specified sequence. For instance, pilots remember to control the individual functions of the plane before flight with the help of a checklist.

There are indications that cooperation scripts as externally induced instructional support must be applied with care. Scripts may be detrimental to collaborative knowledge construction when discussants are more experienced or when scripts are too detailed (Baker & Lund, 1997; Cohen, 1994; Dillenbourg, 2002; Salomon & Globerson, 1989). Cooperation scripts may disturb "natural" interactions and cognitive processes. Cooperation scripts aim at specific activities of collaborative learners, but an a priori structure of discourse cannot foresee any ambiguity or necessary "side tracks" in collaborative knowledge construction. Especially advanced learners may apply specific successful knowledge construction strategies that no typical script may recognize. In particular a very detailed prescription of interactions may hamper knowledge construction. Complex problems may afford a big number of various interactions and allow many solution paths. A detailed structure may, however, reduce the required multiple perspectives on complex problems. Furthermore, less structured learning environments have been argued to better facilitate motivated learning (Reeve, 1996). Cooperation scripts may also overscript novice learners. Dansereau (1988) suggests that cooperation scripts may be directed towards too many aspects and learners may be able to realize only parts of cooperation scripts. For instance, epistemic cooperation scripts may distract learners from referring to each other and social cooperation scripts may suggest that learners disregard the task and carefully plan interactions (Dansereau, 1988, p. 118):

> Some dyads appear to focus on content at the expense of accurately performing the strategy while others do the opposite. In addition, some dyads do not appear to adequately understand and implement the strategy. Finally, some of the dyadic interactions are characterized by a misuse of the available study time through extraneous conversation, concentration on unnecessary details, and occasional arguments.

Thus, differential effects of social and epistemic cooperation scripts on processes and outcomes of collaborative knowledge construction may be expected.

Prompt-based cooperation scripts provide highly detailed structure, because they specify, sequence, and assign roles. Beyond that, these scripts provide prompts to define specific activities for the individual roles. Prompts of CSCL scripts, however, can be flexibly applied by the learners. Some studies show that learners rather ignore CSCL script suggestions (cf. Dillenbourg, 2002; Gräsel, Mandl, Fischer, & Gärtner, 1994; Veerman & Treasure-Jones, 1999). Therefore it can be said that CSCL scripts may not determine any activity of collaborative learners. Instead, learners may choose freely whether or not to apply CSCL scripts. Still, the strategies suggested by cooperation scripts may not be internalized or reduce knowledge acquisition. Larson et al.'s (1985) study indicates that some specialized scripts may actually impede knowledge construction. Larson et al. (1985) argue that scripts may distract learners from learning goals. Scripts may (over-)simplify complex learning tasks which could possibly impede internalization (cf. Reiser, 2002). King (1989a, 1999) shows however, that learners were able to internalize question asking strategies in the guided peer questioning approach to foster individual knowledge acquisition.

Cooperation scripts may therefore need to aim at specific process dimensions to achieve actual facilitation of collaborative knowledge construction. It is important to note that learning can always be constrained by more or less explicit scripts which may either be externally induced by a facilitator or can pose a cultural practice which is already represented in the mind (cf. O'Donnell, 1999; Schank & Abelson, 1977). For instance, it has been well examined, how the tasks learners collaboratively work on affect and structure interaction (Cohen, 1994; Howe, Tolmie, & MacKenzie, 1995). Depending on the task, learners appear to activate specific internally represented scripts. Problem-solving tasks may initiate several internally represented social scripts and provide roles like "superior," who coordinates problem solutions, and "executive," who contributes or executes problem solutions. There are indications, for instance, that in CSCL the social roles of 'thinker' and 'typist' eventually emerge, which produces negative effects on collaborative knowledge construction (Bruhn, 2000). Therefore, externally induced cooperation scripts may need to be designed for specific contexts and with sufficient degrees of freedom for the discussants.

Computer-based media may pose an ideal test bed for adapting the degrees of freedom of cooperation scripts to various contexts and enable rather than constrain interactions (Dillenbourg, 2002). Beyond that, interfaces may be designed based on acknowledged and well researched cooperation script approaches in order to realize instructional support for CSCL.

# 5 Conceptual Framework of the Study and Research Questions

It has been argued that specific *processes* of collaborative knowledge construction foster learning *outcome*. These specific processes hardly emerge spontaneously, but need to be facilitated. Cooperation scripts pose an *instructional approach* that can be oriented towards these specific processes and that can be realized with prompts and implemented in CSCL environments. This study aims to analyze and facilitate collaborative knowledge construction in typical CSCL environments that are based on text-based CMC in web-based discussion boards. In this chapter, the conceptual framework of the study will be summarized with respect to processes, outcomes, and facilitation of collaborative knowledge construction in CSCL environments. Finally, the research questions of this study will be outlined.

## 5.1 The Conceptual Framework of the Study

Based on the theoretical layout of chapter 2, a conceptual framework for the analysis of co-construction processes can be conceptualized. It has been argued that based on socio-constructivist and socio-cultural perspectives, collaborative knowledge construction can be characterized by its social and cognitive processes. These processes may be apparent and well accessible in the discourse of collaborative learners.

Processes of collaborative knowledge construction may be influenced by a number of *context factors* (organizational background, incentive structure, learning task, individual learning prerequisites, cooperation scripts). Some of these context factors are hardly varied in systematic research, but rather have been identified in meta-analyses. Typically, it is difficult to experimentally vary organizational background or learning tasks. The learning environment can be implemented in the standard curriculum of students to foster ecological validity. Thereby, collaborative learning conditions may resemble and be comparable to standard conditions of learners in groups in higher education within the natural variance. Further context factors include individual learning prerequisites which are assessed and tested with respect to the question whether or not experimental randomization of participants of the study was effective (see table 5.1a).

*Table 5.1a:*    Control variables – individual prerequisites for collaborative knowledge construction

| |
|---|
| **Cognitive individual learning prerequisites** |
|     Prior knowledge |
|     Learning strategies |
| **Emotional and motivational individual learning prerequisites** |
|     Social anxiety |
|     Uncertainty orientation |
|     Interest |
| **Computer-specific attitudes** |
|     Attitude towards computers as threat to society |

Thus, all context factors except cooperation scripts are experimentally controlled and are not supposed to substantially differ between the experimental conditions.

Social and cognitive processes have been identified as vital aspects of collaborative knowledge construction. With respect to *social processes*, a conceptual framework of social modes of co-construction, and with respect to *cognitive processes*, a framework of epistemic activities can be applied (see table 5.1b; cf. Fischer, 2001; Fischer et al., 2002).

The *social modes of co-construction* define how learners interpersonally link contributions to each other. Some more transactive modes have been found to trigger more reflective approaches of collaborative learners and to foster collaborative knowledge construction. In the framework of this study, these modes include the externalization and the elicitation of knowledge as well as quick consensus building, integration-oriented consensus building, and conflict-oriented consensus building.

*Table 5.1b*:    Dependent process variables – Fischer's (2001) conceptual framework of the social and cognitive processes of collaborative knowledge construction modified regarding quick consensus building

---

**Social processes: Social modes of co-construction**

    Externalization

    Elicitation

    Quick consensus building

    Integration-oriented consensus building

    Conflict-oriented consensus building

**Cognitive processes: Epistemic activities**

    Construction of problem space

    Construction of conceptual space

    Construction of relations between conceptual and problem space

*Non-epistemic activities*

---

*Epistemic activities* point towards the actual tasks and the contents learners deal with. Generally, epistemic activities have been regarded to as more valuable for knowledge construction than non-epistemic activities. Epistemic activities can be further distinguished into the construction of problem space, the construction of conceptual space, and the construction of relations between conceptual and problem space. It is not well examined, in what relative frequency these activities should appear to foster knowledge construction. Nevertheless it has been argued that the construction of relations between conceptual and problem space may be the pivotal activity of learners in collaborative knowledge construction that is based on complex problems. Furthermore it has been shown that learners who rather construct conceptual space or relations between conceptual and problem space may acquire more knowledge than learners who rather construct problem space.

Chapter 3 was dedicated to the expected *outcomes* of collaborative knowledge construction (see table 5.1c). For this reason, a knowledge-in-use metaphor has been applied. Knowledge-in-use can be regarded to as a co-construct of a group that is displayed in the discourse of learners. Furthermore, learners may acquire knowledge individually. Several types and qualities of knowledge can be distinguished. In the context of collaborative knowledge construction, focused and multi-perspective applicable knowledge can be distinguished as pivotal learning outcomes. Apart from being able to apply theoretical concepts to the elementary sub-problems of a case, collaborative knowledge construction explicitly aims to facilitate the ability to apply knowledge from multiple perspectives.

*Table 5.1c*:   Dependent outcome variables – conceptual framework of outcomes of collaborative knowledge construction

| **Outcomes of collaborative knowledge construction** |
| --- |
| *Applicable knowledge as co-construct* |
| Focused applicable knowledge as co-construct |
| Multi-perspective applicable knowledge as co-construct |
| *Individually acquired applicable knowledge* |
| Individually acquired focused applicable knowledge |
| Individually acquired multi-perspective applicable knowledge |

Next, the question needs to be addressed of how these outcomes may be fostered by process-oriented *instructional support*. In particular, cooperation script approaches that specify, sequence, or assign roles and activities have been outlined. The approach of *CSCL scripts* has been introduced as a combination of the instructional approaches of scripted cooperation and guided peer questioning that can be implemented in CSCL environments. The question remains, however, at what processes in particular CSCL scripts should aim at. Socio-cognitive approaches emphasize the relevance of social and cognitive processes for collaborative knowledge construction. Two scripts have therefore been conceptualized to facilitate either social or epistemic activities. Social and epistemic scripts are expected to facilitate specific social modes and epistemic activities. In this way, social and epistemic scripts can relate explicitly to the portrayed processes and outcomes of collaborative knowledge construction. It is possible that both scripts may facilitate both social and cognitive processes, as well as the outcomes of collaborative knowledge construction (see table 5.1d).

*Table 5.1d*:    Independent variables – instructional support of collaborative knowledge construction with CSCL scripts

**CSCL scripts**

Social cooperation script

Epistemic cooperation script

The *social cooperation script* aims to facilitate specific social modes of co-construction. The social script provides learners with roles and suggests specific forms of interaction that learners rarely engage in spontaneously. The social cooperation script particularly provides the roles case analyst and constructive critic. The function of the case analyst is to start discussion by providing a first analysis of a problem. The tasks of the constructive critics include critical questioning, stating difference of opinions, and providing advice for adjusting the first analysis. The case analyst in turn is expected to respond to these critical questions and comments of the constructive critic, and to author a final, possibly improved analysis of the problem case.

The *epistemic cooperation script* aims to facilitate the specific epistemic activities in collaborative knowledge construction. The epistemic script provides task strategies for learners to endorse the task strategies learners would apply spontaneously, which may be sub-optimal. The task strategy that the epistemic cooperation script suggests is to first collect relevant case information, apply theoretical concepts to the case information, predict possible outcomes of the problem case, and discuss case information, which may not be explained by the theoretical concepts.

Both of these scripts can be realized with prompts in CSCL environments. Learners are expected to respond to specific prompts of the respective scripts. By responding to the prompts in the intended way, the learners follow the task strategies provided by epistemic scripts or fill out the roles suggested by the social script. Furthermore, social scripts can guide learners through the virtual spaces in text-based CMC and suggest learners to contribute to discourse at specified times.

*Table 5.1e*:    Treatment check variables – prompts responded to in the intended way and number of messages

| |
| --- |
| **Prompts** |
| Prompts not responded to |
| Prompts not responded to in the intended way |
| **Structure** |
| Number of messages |
| Heterogeneity of number of messages |

Learners, however, may not follow the respective CSCL scripts and may disregard prompts or not follow a given structure. Therefore a treatment check should clarify to what extent learners have engaged in the activities suggested by the script. First of all, the frequency of prompts that have not been responded to, as well as the frequency of prompts that have not been responded to in the intended way, can be analyzed. Furthermore, the amount of messages suggested by social scripts and the heterogeneity of this number of messages within the learning groups can be analyzed and compared to script prescriptions (see table 5.1e).

## 5.2      Research Questions

Against the theoretical background of socio-constructivist and socio-cultural perspectives and findings on cooperation scripts, social and epistemic cooperation scripts should be investigated in order to identify how collaborative knowledge construction in CSCL environments can be facilitated. Therefore, several research questions can be formulated. *First* it will be examined, to what extent learners applied the CSCL scripts with a treatment check. *Second*, effects of social and epistemic scripts on social and cognitive processes of collaborative knowledge construction will be examined. *Third*, effects of social and epistemic scripts on outcomes of collaborative knowledge construction will be studied. *Fourth*, relations between processes and outcomes of collaborative knowledge construction will be explored. These research questions will be quantitatively pursued. *Finally*, qualitative case studies will be applied to identify and evaluate discourse structures with specific social and cognitive characteristics, and typical comprehension failures.

## 5.2.1 Research questions on facilitation of social and cognitive processes by CSCL scripts

First, the effects of social cooperation script and epistemic cooperation script should be examined in respect to the processes of collaborative knowledge construction.

1a)    To what extent do social cooperation script and epistemic cooperation script and the combination thereof affect processes of collaborative knowledge construction in CSCL environments with respect to **social modes of co-construction**?

There are indications that specific social modes of co-construction can be facilitated with social scripts implemented in a CSCL environment (Nussbaum et al., 2002). Social cooperation scripts facilitate specific social modes of co-construction. As the applied social cooperation script suggests showing elicitation and conflict-oriented consensus building, these social modes should be facilitated.

*Hypotheses*:    The social cooperation script facilitates elicitation.
                The social cooperation script facilitates conflict-oriented consensus building.

Some studies indicate that pedagogical interventions, which aim to structure epistemic activities, also affect social modes of co-construction (Fischer et al., 2002). Epistemic cooperation scripts influence social modes. As the applied epistemic cooperation script suggests to give a more detailed

81

analysis of the problems, learners may engage more in the social mode of externalization.

*Hypothesis*: The epistemic cooperation script facilitates externalization.

1b)   To what extent do social cooperation script and epistemic co-operation script and the combination thereof affect processes of collaborative knowledge construction in CSCL environments with respect to **epistemic activities**?

Empirical findings point out that structuring CSCL facilitates epistemic over non-epistemic activities (e.g., Baker & Lund, 1997). This has been confirmed, even though CSCL without any particular structure is already supporting epistemic activities compared to FTF discourse (Woodruff, 1995). Furthermore, empirical studies show that specific epistemic activities can be facilitated in CSCL (Fischer et al., 2002). Therefore, learning groups with any form of CSCL script engage in less non-epistemic and more epistemic activities than unscripted learning groups. Furthermore, an epistemic CSCL script facilitates specific epistemic activities. The epistemic cooperation script particularly aims at the construction of problem space in order not to mimic expertlike approaches, but to limit concrete operations within the frame of specific prompts. Furthermore, the epistemic cooperation script aims to facilitate the construction of relations between conceptual and problem space. Therefore, these epistemic activities may be facilitated with the epistemic cooperation script. No main effects or interaction effects with social cooperation scripts on specific epistemic activities are expected, however.

*Hypotheses*:   Both social and epistemic cooperation scripts reduce the frequency of non-epistemic activities.
The epistemic cooperation script facilitates construction of problem space.
The epistemic cooperation script facilitates construction of relations between conceptual and problem space.

### 5.2.2 Research questions on facilitation of learning outcomes by CSCL scripts

Second, effects of social cooperation script and epistemic cooperation script should be examined with respect to outcomes of collaborative knowledge construction. Based on socio-constructivist and socio-cultural approaches, scripts facilitating specific processes should also facilitate specific outcomes of collaborative knowledge construction.

2a) To what extent do social cooperation script and epistemic cooperation script and the combination thereof affect outcomes of collaborative knowledge construction in CSCL environments with respect to **focused and multi-perspective applicable knowledge as co-construct**?

*Hypothesis*: Both social and epistemic cooperation script foster focused and multi-perspective applicable knowledge as co-construct.

2b) To what extent do social cooperation script and epistemic cooperation script and the combination thereof affect outcomes of collaborative knowledge construction in CSCL environments with respect to **individual acquisition of focused and multi-perspective applicable knowledge**?

*Hypothesis*: Both social and epistemic cooperation scripts facilitate individual acquisition of focused and multi-perspective applicable knowledge.

### 5.2.3 Research questions on relations of processes with outcomes of collaborative knowledge construction

Finally, relations of variables examined during the collaborative phase (social processes, cognitive processes, and knowledge as co-construct) and individual acquisition of applicable knowledge will be examined exploratory. These relations should be strong according to both socio-constructivist and socio-cultural approaches. Based on socio-cultural approaches, learners should be able to apply knowledge individually that they have applied in prior collaborative phases. Thus, there should be a strong relation between applicable knowledge as co-construct and individual acquisition of applicable knowledge.

3a)  To what extent are **social processes** related to the individual acquisition of applicable knowledge?

*Hypothesis*: Social processes relate strongly with the individual acquisition of applicable knowledge.

3b)  To what extent are **cognitive processes** related to the individual acquisition of applicable knowledge?

*Hypothesis*: Cognitive processes relate strongly with the individual acquisition of applicable knowledge.

3c)  To what extent is the **applicable knowledge as co-construct** related to the individually acquired applicable knowledge?

*Hypothesis*: Knowledge as co-construct relates strongly with the individual acquisition of applicable knowledge.

### 5.2.4      Research questions of the case studies

The quantitative analyses record frequencies of specific social and cognitive processes. Qualitative analyses may further reveal how specific processes affect collaborative knowledge construction. Therefore, qualitative case studies can illustrate the quantitative results. In particular, the case studies aim to answer the following research questions.

4a)  Which social and cognitive discourse structures can be identified in the experimental conditions?

4b)  Which comprehension failures of students can be identified?

# 6      Methods of the Empirical Study

In this chapter, sample and design, the experimental learning environment, and the variables will be reported. Apart from the statistic procedures, the qualitative approach of a graphical coding analysis will be introduced.

## 6.1      Sample and Design

96 students of Educational Sciences from the Ludwig-Maximilians-Universität (LMU) München participated in this study. The students, who were attending a mandatory introduction course, participated in an online learning session as a substitute for one regular session of the course. Participation was required in order to receive a course credit at the end of the semester. The learning outcomes of the experimental session, however, were not accounted for in students' overall performance. The introduction course consists of a one hour lecture and a successive two hour seminar. Equally, the online learning session took three hours.

The data was collected in 32 separate experiments distributed over a period of two months (January and February 2001) with groups of three students. The students were to learn Weiner's (1985) attribution theory in the online learning session, which is standard curriculum content of the introductory course. Students were individually invited to one of three different laboratory rooms. Learning partners did not meet or know each other before the experimental session. Each group was randomly assigned to one of the four experimental conditions in a 2×2 factorial design (see figure 6.1). The four experimental conditions will be abbreviated with DWS (= discourse without script), ECOS (= epistemic cooperation script), SCOS (=

social cooperation script), and ESCOS (= epistemic and social cooperation script).

|  | | Social cooperation script | |
|---|---|---|---|
|  | | Without | With |
| **Epistemic cooperation script** | Without | $n = 8$ | $n = 8$ |
|  | With | $n = 8$ | $n = 8$ |

*Figure 6.1*:  2×2 factorial design of the empirical study with eight groups of three equaling 24 participants in each experimental condition.

The factors "social cooperation script" (none vs. social cooperation script) and "epistemic cooperation script" (none vs. epistemic cooperation script) were experimentally varied. Data was collected in a pre-test, during the treatment phase, and in an equivalent post-test (see section 6.6 for a detailed description of the control variables).

*Table 6.1a*:  Demographic data of the participants in the four experimental conditions.

|  | DWS | ECOS | SCOS | ESCOS |
|---|---|---|---|---|
| **Gender** | | | | |
| - Female | 21 | 20 | 23 | 19 |
| - Male | 3 | 4 | 1 | 5 |
| **Age** | $M = 23$ $(SD = 5)$ | $M = 22$ $(SD = 3)$ | $M = 22$ $(SD = 3)$ | $M = 23$ $(SD = 4)$ |
| **First language** | | | | |
| - German | 19 | 17 | 20 | 18 |
| - Other | 5 | 7 | 4 | 6 |

The sample can be described as follows (see table 6.1a): 83 participants were female and 13 were male. The age of the participants averaged at 23 years (SD = 4). 74 participants declared German as their first language. It should also be noted that the 22 students whose first language was not German have spoken German for an average of 10 years (SD = 7). The experimental groups did not differ systematically with respect to gender, age, or first language.

*Table 6.1b*:   Prior applicable knowledge of the participants in the four experimental conditions.

|  | DWS | ECOS | SCOS | ESCOS |
|---|---|---|---|---|
|  | *M*  (*SD*) | *M*  (*SD*) | *M*  (*SD*) | *M*  (*SD*) |
| **Prior applicable knowledge** | | | | |
| Focused | 0.13 (0.61) | 0.04 (0.20) | 0.29 (0.81) | 0.21 (0.51) |
| Multi-perspective | 0.42 (0.78) | 0.38 (0.88) | 0.25 (0.44) | 0.33 (0.76) |

The experimental groups did not differ systematically with respect to prior applicable knowledge (see table 6.1b). The university students disposed of very little prior knowledge in general. Therefore, the portrayed differences are subject to a floor effect and cannot be reliably measured (see section 6.6).

*Table 6.1c*:   Learning prerequisites of the participants in the four experimental conditions.

|  | DWS | ECOS | SCOS | ESCOS |
|---|---|---|---|---|
|  | *M*  (*SD*) | *M*  (*SD*) | *M*  (*SD*) | *M*  (*SD*) |
| **Learning strategies** | 3.19 (0.51) | 3.11 (0.46) | 3.21 (0.68) | 3.29 (0.44) |
| **Social anxiety** | 2.23 (0.92) | 2.24 (0.82) | 2.51 (1.04) | 2.32 (0.73) |
| **Uncertainty orientation** | 3.28 (0.52) | 3.32 (0.61) | 3.32 (0.66) | 3.32 (0.48) |
| **Computer-specific attitudes** | 3.00 (0.69) | 2.98 (0.68) | 2.92 (0.70) | 2.90 (0.62) |
| **Interest** | 4.12 (0.69) | 4.06 (0.84) | 3.93 (0.84) | 3.76 (0.73) |

Furthermore, the randomization of the four experimental groups was effective with respect to important prerequisites of collaborative knowledge construction like learning strategies, social anxiety, uncertainty orientation, computer-specific attitudes, and interest towards the learning environment (see table 6.1c). All of these control variables were measured with individual instruments, described in section 6.6, that are all based on a five-point Likert scale ranging from low (= 1) to high (= 5).

## 6.2     Experimental Phases

*Table 6.2:*     Overview of the test procedure

| | Duration |
|---|---|
| **(1) Introduction and pre-tests** | |
| Introductory explanations | 5 min |
| Assessment of learning prerequisites (questionnaire) | 5 min |
| Pre-test of applicable knowledge (case) | 10 min |
| **(2) Individual learning phase** | |
| Introductory remarks | 5 min |
| Individual study phase of the theoretical text | 15 min |
| **(3) Collaborative learning phase** | |
| Introduction to the technical handling of the web-based learning environment | 20 min |
| Explanation of the procedure | 5 min |
| Collaborative learning phase | 80 min |
| **(4) Post-tests and debriefing** | |
| Post-test of applicable knowledge (case) | 10 min |
| Assessment of the subjective learning experience | 10 min |
| Debriefing | 5 min |
| **Total time** | ca. 3 h |

The experiment extended over four phases (see table 6.2). (1) *Introduction and pre-test*: first, the participants of the study filled out a questionnaire and a pre-test consisting of a case task. (2) *Individual learning phase*: after the pre-test, the students were asked to individually study a three page description of the attribution theory. (3) *Collaborative learning phase*: after the learners were briefly introduced to the handling of the learning environment, they worked together on three cases. In this collaborative phase the individual experimental groups were provided with the individual treatments. (4) *Post-tests and debriefing*: the collaboration was followed by an individual post-test which paralleled the individual pre-test.

The time in the individual phases was exactly the same in all four conditions. In the following paragraphs, the individual phases of the test procedure will be outlined in more detail. Finally, training of the experimenters who surveyed the four phases of the experiment will be described.

### 6.2.1 Introduction and pre-tests

First of all, a possible field of application of the results of the empirical study was pointed out to the participants. The possible field of application of the results of the empirical study was to improve learning with new media at the university level, e.g., at the Virtual University of Bavaria. Furthermore, the university students were introduced to the learning goals of the experimental web-based environment. These learning goals were to experience forms of virtual learning with new media, which represents an important focal point of the studies of Educational Sciences at the LMU, and to learn about a prominent theory of Educational Sciences together with two learning partners. The participants were assured that all data were treated anonymously. The participants were informed that they were surveyed via video, but were further assured that they were not recorded. After that, the participants were equipped with a pen, paper, and a marker for making notes. Furthermore, learners were provided with a clock to monitor time in the individual segments of the experiment. Finally, the participants were asked to fill out a questionnaire of some demographic and other control variables (e.g., uncertainty orientation) and analyze a baseline case to their best knowledge (see section 6.6).

### 6.2.2 Individual learning phase

In the individual learning phase, the participants were informed about the general procedure of the online learning session, which entailed analyzing a few more problem cases with learning partners mediated via text-based communication "similar to communication with email or SMS." Subsequently, the participants were handed out a theoretical text about attribution theory (see section 6.3.1) and were told that their learning partners would receive the same theoretical text. Finally, the experimenters pointed out once more to the students that they could make notes and mark the text.

### 6.2.3 Collaborative learning phase

In the individual laboratory rooms, each student was equipped with a standard IBM Pentium computer with a standard web-browser (MS Internet Explorer). With the help of this hard- and software, the students could communicate with each other via web-based discussion boards of the learning environment (see figure 6.2.3).

**Laboratory room 1**      **Laboratory room 2**      **Laboratory room 3**

*Figure 6.2.3*:  A learning group of three participants in separate rooms communicating via web-based discussion boards.

First of all, the participants were introduced to the handling of the learning environment. They were given socio-emotionally neutral code names (Ahorn, Birke, or Pinie), and were informed about the individual components of the web-based learning environment (timer, task description, orientation map, and web-based discussion board; cf. section 6.3.3). The participants were asked to write and read some of the messages in test boards. They were taught how to title messages, start threads, answer messages, quote and delete text in the text windows, and read the messages of the various levels of a thread (initiating message, response etc.). Furthermore, the participants were introduced to the specifics of the individual experimental conditions (cf. section 6.4). The task of the students was to discuss three problem cases on grounds of the theoretical text and to come to an agreement about an analysis of each of the three cases. Finally, the participants were asked to read the 'Welcome'-page of the learning environment (see figure 6.3.3a) and wait until all learning partners were ready to start at the same time. From this moment on, the learning environment worked automatically. During collaboration, all participants had a copy of the text covering attribution theory. The whole discourse of the students was recorded by means of web-based discussion boards through which the participants communicated. A quarter of an hour prior to the end of the collaborative phase, the students were reminded by the experimenters to fulfill the task and come to joint analyses.

### 6.2.4 Post-tests and debriefing

After the collaborative phase, the participants analyzed another problem case individually with pen and paper. The procedure was equivalent with the baseline case of the pre-test. Finally, in a debriefing interview, the participants were asked to report their experiences in general. These comments were noted by the experimenter.

### 6.2.5 Experimenter Training and Surveillance of the Experiment

The four general phases of the experiment were introduced by intensively trained experimenters who were equipped with a detailed portfolio of each of the four experimental conditions. The training of the experimenters included the accomplishment of the experiment in the role of the subjects, extensive practical training in the role of experimenter assistants, and role playing as responsible experimenters in various worst case scenarios (e.g., system crash). Furthermore, the experimenters were trained in how to respond to questions posed by participants.

*Figure 6.2.5*: Observation of the participants and their computer monitors.

The four phases of the experiment were surveyed by video. The participants themselves as well as their computer monitors were observed from a video control room (see figure 6.2.5). The surveillance of the experiment via video served to reduce eventual experimenter effects and to better control the procedure of the experiment. The participants were not recorded on tape.


## 6.3　　　Learning Material and CSCL Environment

This section serves to illustrate the learning material as well as the technical aspects of the web-based learning environment. This includes the theoretical text and the problem cases as well as the individual components of the web-based learning environment.

### 6.3.1 Theoretical text

The task given to the students was to analyze cases with the help of Weiner's (1985) attribution theory. First of all, students were handed a short description of the attribution theory, which they were allowed to study on their own (including taking notes and marking text). The theoretical text mainly covered Weiner's (1985; Weiner et al., 1971) attribution theory and addresses the question how students attribute causes for success or failure. Weiner's work is strongly based on the conceptions of Heider (1958) who allocates causes for attribution to two dimensions, namely locality and stability.

The dimension of *locality* means that attributed causes can be found within or outside of a person who experienced success or failure. In other words, causes for success or failure can be attributed to internal or external causes. An example of an internal cause would be the time the individual learner invested prior to an exam. An external cause is, for example, the difficulty of an exam. Regarding the dimension of *stability*, attributed causes may be temporally stable or variable. This dimension describes whether attributed causes may be different next time, because they vary by chance or because they could be changed intentionally, as for example, when a student decides to invest more time in preparing for an exam. In contrast, an example for a stable cause is the talent of a student, which is a relatively constant (stable) feature. These two dimensions are independent of each other and thus pose a matrix of four attribution variants: Talent, effort, task difficulty, and chance (see figure 6.3.1).

There is a prescriptive aspect to this classification system. Weiner (1985) assumes that in order to sustain learning motivation, failures should be attributed to variable causes such as chance or effort. Students who ascribe failures to a lack of effort may work harder in the future, whereas students who ascribe failures to a lack of talent may regard further learning efforts as pointless, because they believe they cannot improve or change anything in comparable future situations. Accordingly, attributions of success to talent are beneficial for future performance, whereas attributions of success to chance are detrimental.

| | | Locality | |
|---|---|---|---|
| | | Internal | External |
| **Stability** | Stable | Talent | Task difficulty |
| | Variable | Effort | Chance |

*Figure 6.3.1*: Classification system of attributions according to Heider (1958)

The attribution theory further distinguishes between attributions of the concerned student him- or herself and other persons, which can have equivalent effects on future performance. The short text about the attribution theory concludes with a reference to re-attribution-training, which may change detrimental and foster beneficial attributions (Ziegler & Heller, 2000).

### 6.3.2    Problem cases

After studying the theoretical text, the task of the students was to analyze and discuss problem cases. An important aspect of learning based on problems is that the problems that learners are confronted with, should be complex and require the construction of problem space, conceptual space, and the construction of relations between conceptual and problem space (De Jong & van Joolingen, 1998; Fischer et al., 2002; Kitchner, 1983). Thus, presented problem information can be irrelevant or ambiguous. These complex problems are supposed to resemble real world contexts familiar to students, which may facilitate motivation and knowledge construction in complex domains. In this regard, the presentation format of problems may be of importance. Some instructional approaches, for example, use simulations or a videodisc format to present problems in a more realistic way (*Anchored Instruction*, Cognition and Technology Group at Vanderbilt, 1997). However, these formats may be rather costly. The format of text-based cases is a more feasible approach to present problems and is well known in juristic, management, and medical education, as well as in problem-oriented approaches to learning (Kaiser, 1983).

With respect to the learning environment of the study, problem cases were developed that resembled daily situations of students and that could be solved with the help of the attribution theory. For instance, learners were supposed to analyze the following problem case:

> You participate in a school counseling as a student teacher of a high school with Michael Peters, a pupil in the 10th grade.
>
> "Somehow I begin to realize that math is not my kind of thing. Last year I almost failed math. Ms Weber, who is my math teacher, told me that I really had to make an effort if I wanted to pass 10th grade. Actually, my parents stayed pretty calm when I told them. Well, mom said that none of us is 'witty' in math. My father just grinned. Then he told that story when he just barely made his final math exams with lots of copying and cheat slips. 'The Peters family,' Daddy said then, 'has always meant horror to any math teacher.' Slightly cockeyed at a school party, I once have told this story to Ms Weber. She said that this was no bad excuse, but no good one either. Just an excuse that is, and you could come up with some more to justify to be bone idle. Last year I have barely made it, but I am really anxious about the new school year!"

The descriptions of the problem cases were embedded into the web-based learning environment, so that the participants could study the problem case while authoring new messages on the web-based discussion boards.

### 6.3.3 Web-based learning environment

The web-based learning environment is a password protected website in which three participants can post messages that, apart from the experimenters, only the members of the learning group could read. The participants were logged in with code names in an effort to warrant anonymity. The learning group was locally spread out, which means that the individual participants operated on separate, distant computer terminals. The web-based environment consists of several components that are based on html, php- and perl-protocols, and javascript (Stegmann, 2002). The various modifiable components of the learning environment will be described in the following paragraphs.

### Html-web-pages

The whole web-based learning environment is accessible via the World Wide Web and built on standard html-web-pages. Html (Hypertext markup language) is a format that may combine text and images and is readable with browsers of any operating system or computer model (see figure 6.3.3a).

*Figure 6.3.3a*: The 'Welcome!'-web-page of the learning environment with a summary of the task combining text and images.

### Flexible task description and timer (javascript)

One component of the web-based learning environment is a flexible task description and a timer in a dark green window which is continually present in the upper left corner of the screen (see figure 6.3.3b). These features are based on javascript and may be modified and adopted to different settings of the learning environment. For instance, the timer counts down the minutes of the task. The timer can also structure the collaborative phase into several time segments. This helps to guide learners through a series of interactions in various web-based discussion boards. Accordingly, the task description can be adjusted to any of the programmed segments. In other words, any time segment can be characterized by a different task description.

*Figure 6.3.3b*: Flexible task description and timer.

### Web-based discussion boards (php- and perl-protocols)

The main component of the communication interface consists of an arbitrary number of web-based discussion boards. Each board has a distinct color and a map in the lower left corner to facilitate orientation (see figure 6.3.3c). The orientation map marks the current discussion board with a red ×. On the boards, a description of the problem cases can be found. Below the case description, a text message (incl. a header) can be typed into text windows. It is possible, to pre-structure these text windows with prompts to which learners can flexibly react to.

After contributions have been sent off, they can be accessed via an overview page of the individual web-based discussion boards (see figure 6.3.3d). In this overview page, the typical thread structure of discussions is shown. A specific thread structure can be realized with the help of the timer. The timer may guide participants in a specific sequence through the web-based discussion boards. This sequence can be visualized with the help of the orientation map that can indicate a specific sequence of the individual web-based discussion boards with arrows.

Task descrip-
tion and timer

Problem case
description

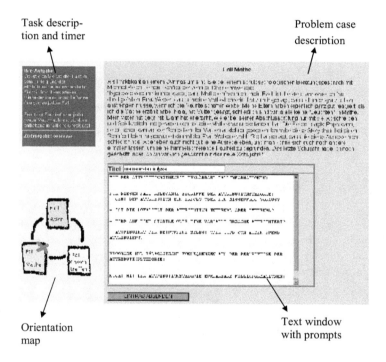

Orientation
map

Text window
with prompts

*Figure 6.3.3c*: Interface of the learning environment with an orientation map, the case description, and text-window (incl. prompts).

*Figure 6.3.3d*: Overview of one out of three completed web-based discussions that was guided and segmented with the help of a timer and an orientation map.

## 6.4        Experimental Conditions

In this section, the four experimental conditions of the 2×2-design will be described. In this context, the implementation of the social cooperation script and the epistemic cooperation script into the web-based learning environment will be illustrated.

### 6.4.1        Discourse without cooperation scripts (DWS)

The participants of the experimental condition "Discourse without cooperation scripts" (DWS) were allowed to access three distinct web-based discussion boards within the web-site via an overview page (see figure 6.4.1). The problem case descriptions were permanently available inside the individual discussion boards. Within these discussion boards new contributions (initiating messages) could be posted that started a thread or existing messages could be answered in order to continue a thread with an unlimited number of responses.

*Figure 6.4.1*:  Overview-page of all three discussion boards for the problem
cases

As in standard web-based discussion boards, replies to messages
contain the original text of the message which is marked with ">." Additionally, quoted text is presented in different colors to be able to more easily
distinguish between quoted text and new words. The participants were introduced to the possibility of the web-based discussion board to delete
original text when composing a new message in the text window. The timer
in the DWS-condition was set to 80 minutes after which the collaborative
phase automatically ended.

### 6.4.2      Social cooperation script (SCOS)

Each student involved in the "social cooperation script" scenario
(SCOS) was assigned two roles: (a) analyst for one of the cases and (b) constructive critic for the other two cases. Role (a) included taking over the
responsibility for the preliminary and concluding analysis as one case and
responding to criticism from the learning partners. In their function of a
constructive critic (role (b)), the learners were required to criticize the
analyses of the two other cases presented by the learning partners. These
activities were supported by the interaction-oriented prompts (see figure
6.4.2a), which were automatically inserted into the critics' messages and
into the analyst's replies in order to help learners successfully take over
their roles.

**Prompts for the constructive critic**

These aspects are not clear to me yet:

We have not reached consensus concerning these aspects:

My proposal for an adjustment of the analysis is:

**Prompts for the case analyst**

Regarding the desire for clarity:

Regarding our difference of opinions:

Regarding the modification proposals:

*Figure 6.4.2a*: Prompts of the social cooperation script to support the roles of constructive critic and case analyst in collaborative knowledge construction.

SCOS-learners were given a time limit for each of the required activities. All in all, these activities lasted 80 minutes, as in the groups without the script. The SCOS-learners were guided through all three cases and were asked to alternately play the role of the analyst and that of the critic. SCOS-learners were given a list of the respective prompts with short explanations, and a plan of how the social cooperation script would guide the collaborative learning phase (see figure 6.4.2b).

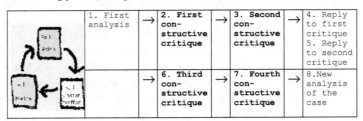

| | 1. First analysis | → | 2. First con-structive critique | → | 3. Second con-structive critique | → | 4. Reply to first critique 5. Reply to second critique |
|---|---|---|---|---|---|---|---|
| | | → | 6. Third con-structive critique | → | 7. Fourth con-structive critique | → | 8.New analysis of the case |

*Figure 6.4.2b*: Plan of the automatic guidance of the social cooperation script through the web-based discussion boards handed out to the participants.

The social cooperation script determined the number of messages being produced for each case (see table 6.4.2; see also figure 6.3.3d for an example of a complete discourse that was SCOS-guided).

All learners had to draft a first analysis for one of the three problem cases (1st message), continue with the second problem case and write a critique of an initial analysis of one of the learning partners (2nd message), continue with the third problem case and write another critique of an initial analysis of the other learning partner (3rd message). The learners then had to return to the original problem case and compose two replies to the two critiques the learning partners have written in the meantime (4th and 5th message), then continue with the second problem case and respond critically to the defense of the analyst of the second problem case (6th message), move on to the third problem case and do the same (7th message), then return to the original problem case and compose a final analysis (8th message).

*Table 6.4.2*:   Result of the cooperation script for one of the three cases
with the respective time limits

| Student A (analyst) | Student B (critic) | Student C (critic) |
| --- | --- | --- |
| First analysis (16 minutes) | | |
| | Constructive critique* (8 minutes) | Constructive critique* (8 minutes) |
| Replies to both critics* (16 minutes) | | |
| | Constructive critique* (8 minutes) | Constructive critique* (8 minutes) |
| Final analysis (16 minutes) | | |

* these contributions were facilitated with prompts for critics and analysts

In summary, the SCOS suggested that learners write eight messages for each problem case: an initial analysis of the case, four critiques in total, two replies, and another, final analysis (see table 6.4.2). In addition to the prompts within the text windows, the SCOS also set the titles of these eight messages.

### 6.4.3    Epistemic cooperation script (ECOS)

Participants of the experimental condition "epistemic cooperation script" (ECOS) disposed of the same resources and technique as the DWS-learners, but any initiating message was pre-structured with ECOS-prompts (see figure 6.4.3).

---

Case information, which can be explained with the attribution theory

Relevant terms of the attribution theory for this case:

-   Does a success or a failure precede this attribution?

-   Is the attribution located internally or externally?

-   Is the cause for the attribution stable or variable?

-   Does the concerned person attribute himself/herself or does another person attribute him/her?

Prognosis and consequences from the perspective of the attribution theory:

Case information which cannot be explained with the attribution theory:

---

*Figure 6.4.3*:   Prompts of the epistemic cooperation script to apply the concepts of Weiner's (1985) attribution theory to problem cases.

The prompts were questions about the problem cases and aimed to facilitate learners to first identify relevant case information, relate the concepts of the attribution theory to the case information, predict outcomes and suggest pedagogical interventions for the case, and finally identify case information which cannot be explained with attribution theory. These prompts were introduced to the ECOS-learners on a piece of paper with short explanations of the meaning of the individual prompts. This prompt list was handed out to the ECOS-learners prior to the collaborative treatment phase while the technical handling of the web-based learning environment was explained. These prompts were supposed to be responded to in the given sequence. For this reason, the prompts were separated from each other by empty lines in which the participants were supposed to type in their respective responses to the prompts. Thus, the specific sub-tasks of the learners were to respond to the given prompts and elaborate on them together with their learning partners. As the original text is quoted in any subsequent message (response), the prompts continue to be present within a discussion thread. As in the DWS-condition, the timer in the ECOS-condition was set to 80 minutes after which the collaborative phase automatically ended.

### 6.4.4 Epistemic and social cooperation script (ESCOS)

In combination of both epistemic and social cooperation scripts (ESCOS), all initiating messages were pre-structured with the ECOS-prompts. The distribution of the roles, including the social prompts as well as the timer-controlled guidance through the three case boards, was identical to the SCOS-condition. In other words, the first and the concluding messages of the analyst were pre-structured with the prompts of the epistemic cooperation script and the application flow was otherwise identical to the SCOS-condition. The participants were given both handouts about the ECOS-prompts and the SCOS-prompts. The time on task was 80 minutes, which was identical to any other experimental condition.

## 6.5 Operationalization of the Dependent Variables

The goal of the empirical study was to measure the effects of the social and epistemic cooperation scripts – the independent variables – on the processes and outcomes of collaborative knowledge construction, which are the dependent variables. In this section all dependent variables, including social and cognitive processes of collaborative knowledge construction, and learning outcomes will be presented. For this reason, process data as well as outcome data have been collected and measured.

## 6.5.1    Process data and their measurement

The data for the social and cognitive processes were collected during the collaborative learning phase. The data source for the process measures is the discourse of one of the three web-based discussion boards. Social modes of co-construction and epistemic activities have been analyzed with a coding system for multi-level analysis of knowledge co-construction processes in discourse (Weinberger & Fischer, 2006; Weinberger, Fischer, & Mandl, 2002), which is based on a coding system by Bruhn, Gräsel, Fischer, and Mandl (1997) that has been applied by Fischer et al. (2002).

### Unit of analysis

In the quantitative analyses, the *unit of analysis* of the process measures is the relationship between two subsequent conceptual components of the discourse of the students (cf. Hofer & Pikowsky, 1993). For instance, the sentence "The cue that no one in the family is witty anyway is equivalent to an attribution on talent" consists of the case information "that no one in the family is witty" on the one hand and the concept "attribution on talent" on the other hand. The relation between the conceptual components is defined by the words "is equivalent." Conceptual components may be spread over syntactical units such as sentences. For instance, "My first impression is that the pupil stresses his lack of talent. This is an internal attribution." One conceptual component is "that the pupil stresses his lack of talent" which is related to another conceptual component, namely "internal attribution." The utterance that relates two conceptual components may be more relevant for collaborative knowledge construction because the coarser granularity of this unit of analysis may indicate components of mental models better than syntactical units (cf. Chi, 1997; Chi, DeLeeuw, Chiu, & La-Vancher, 1994; Hofer & Pikowsky, 1993).

### Rater training

Two raters have been trained to identify these units of analysis by processing ca. 1000 units together. After that, inter-rater reliability was determined on the grounds of roughly 500 units more that the raters analyzed independently from each other. Discrepancies were not resolved in accordance with (Chi, 1997), "because resolving them can actually bias the interpretation of subsequent codings" (p. 307). Coder-correspondence for identifying units was 87%, determined on the grounds of the ca. 500 units, which is an acceptable ratio (cf. Chi, 1997).

### Overall raw data

In total, the discourses of all groups in one of the three web-based discussion boards consisted of 17863 words (not including the prompts) distributed over 412 messages. On average, 184 words (SD = 133.18) were posted in 4 messages (SD = 3.66) per person in one of the three web-based discussion boards. These discourses have been segmented into a total of 1763 codeable units (6 utterances were incomplete and could not be coded). On average, each learner produced 18 codeable units for this one problem case (SD = 14.34). There are no substantial differences regarding the number of units of analysis between the experimental groups. For a clearer representation and better comparability, the process measures of social modes of co-construction and epistemic activities will be additionally reported in percentages. Effects of the scripts will be presented with z-scores.

### Categorization

With the help of the 'Coding System for a Multi-Level Analysis of Knowledge Co-Construction' (Weinberger & Fischer, 2006; Weinberger et al., 2002), each unit can be analyzed simultaneously on levels of social processes (social modes of co-construction) and cognitive processes (epistemic activities).

(1) *Social modes of co-construction.* Any unit of analysis in the discourse was coded with respect to social modes of co-construction that indicate an increasing degree of transactivity, namely, externalisation, elicitation, quick consensus building, integration-oriented consensus building, and conflict-oriented consensus building. The inter-rater reliability regarding social modes of co-construction, which was measured with Cohen's Kappa, was $\kappa = .81$.

(a) *Externalization.* Units have been coded as externalization if they were neither prompted by learning partners or constructed as a response to any other contribution of learning partners. Thus, new initiating messages typically contained externalization, as for instance in "My first impression is that the pupil ascribes his bad performance in mathematics to talent."

(b) *Elicitation.* A unit is coded as elicitation if a learner aims to directly trigger a specific reaction from the learning partners. This is typically done by question asking (e.g., "Why do you believe that Michael ascribes his deficits to an internal, stable cause?"), but syntactically not restricted to questions. Equally, directive discourse moves, such as "You should mention the attribution of the parents," aim to elicit a specific reaction by a learning partner and are accordingly coded as elicitation.

(c) *Quick consensus building* has been coded when learners agree without further comments (e.g., "Yes"). Agreement may be expressed by short sign of approvals or by literal repetition of what has already been said (A: "Michael does not feel like making an effort in math;" B: "He simply does not learn for this subject") or by juxtaposition ("Both are somehow correct").

(d) *Integration-oriented consensus building.* Segments in which contributions by a learning partner are adopted into one's own considerations, but were not previously considered, have been coded as integration-oriented consensus building. This includes sublation of perspectives and explicit adoption of perspectives (e.g., A: "Michael attributes to internal, stable causes;" B: "The parents attribute to talent;" C: "Both Michael and his parents have a detrimental attribution pattern"), but does not refer to mere agreement.

(e) *Conflict-oriented consensus building.* Utterances that dismiss, modify, or devaluate contributions of learning partners have been coded as conflict-oriented consensus building. In this way, not only explicit, but also implicit rejections indicated by slight repairs or modifications of contributions of partners are regarded to as conflict-oriented. Thus, individual conflict-oriented segments are indicated by explicit rejections ("I don't think so"), replacements (A: "The attribution of the teacher is de-motivating;" B: "The attribution of the teacher is beneficial"), modification (A: "The attribution of the parents is positive because it liberates Michael of his feelings of guilt;" B: "It is positive in the sense that the parents do not put pressure on Michael, but accept him principally"), or critical endorsement (A: "The teacher motivates Michael by ascribing his bad performance to laziness;" B: "The teacher motivates Michael by also evaluating the attributions of his parents").

(2) *Epistemic and non-epistemic activities.* Raters had to differentiate the various epistemic activities and if a unit of analysis was epistemic or non-epistemic. The inter-rater reliability of the individual epistemic and non-epistemic activities that were measured with Cohen's Kappa was $\kappa = .90$.

Contributions have been classified as *non-epistemic*, when the content of the contributions was neither about theoretical concepts, nor case information. Thus, typical for non-epistemic activities were students' digressions off topic, e.g., "The weather could be better today." Coordinating activities of learners have been also coded as non-epistemic, e.g., "Where is everybody?". The *epistemic activities* describe the tasks of learners to construct problem space, conceptual space, and interrelations between both spaces.

(a) *Construction of problem space.* Relations between two conceptual components have been coded as construction of problem space when both components are case information. In other words, participants constructed problem space whenever they repeated or paraphrased case information in order to acquire a better model of the situation. One characteristic of the construction of problem space is that no theoretical concepts are reported as in, for example "Michael's mother says that no one in the family is actually witty anyway; this she said with a smile."

(b) *Construction of conceptual space.* If participants built relations between two theoretical concepts, this relation was coded as construction of conceptual space. Thus, the repetition of concepts of the theoretical text has been coded as *construction of theoretical conceptual space.* In the example "An attribution towards talent is an internal, stable attribution," a relation between two theoretical concepts is constructed ("attribution towards talent" and "internal, stable attribution").

(c) *Construction of relations between conceptual and problem space.* When the participants related theoretical concepts and case information, these utterances were classified as constructions of relations between conceptual and problem space. In other words, this category describes how learners applied concepts to problems. For example, in the utterance "That Michael said, 'I am just not talented' points to an internal attribution," the theoretical concept "internal attribution" was applied to the case information "Michael said 'I am just not talented.'"

### 6.5.2    Learning outcome measures

The co-construct as well as the individual acquisition of various forms of knowledge have been considered as learning outcomes. These two dimensions of learning outcomes have been assessed at different times during the experiment. Basis for knowledge as co-construct is the discourse of learning groups. The individual acquisition of applicable knowledge has been assessed in individual post-tests. The data source for the individual acquisition of *applicable knowledge* was the individual analysis of a transfer case. The pen and paper analyses of the individual participants was segmented and coded in the same way as discourse with the 'coding system for a multi-level analysis of knowledge co-construction.' The dimension of social modes of co-construction were not applied, because analysis of the transfer case was accomplished without references to learning partners. 958

units of analysis have been segmented in all analyses of the transfer case. Raters were blind to the experimental condition during the coding procedure of the individual post-test.

In the collaborative learning phase as well as in the individual post-test phase, learners were asked to analyze cases with the help of the attribution theory. Several concepts of attribution theory needed to be connected to information of the individual problem cases. Applicable knowledge has been measured with the amount of adequate relations between theoretical concepts and case information. Individual relations of theoretical concepts to case information have been identified as adequate with respect to the theoretical text and an expert solution, and allocated to *focused* respectively *multi-perspective applicable knowledge*. The problem cases typically raise the question, how a person may perform in future exams. On the one hand, these problem cases contain information that denotes future failures because a person shows some attribution behavior detrimental to learning. These aspects can be regarded to as the most prominent within the problem case or as the core problem. The contributions that built adequate relations of concepts of the theoretical text to these problems have been coded as *focused applicable knowledge*. On the other hand, the problem cases typically contain some information that indicate possibilities for future improvement of the protagonist due to some beneficial attributions of his or her social environment. These attributions of others may counter the most prominent detrimental attributions of the protagonist of the problem case. Thus, a complete analysis of the problem case, in comparison to an expert analysis, must build on a range of aspects of the problem case that may be regarded to as sub-problems. Utterances have been coded as *multi-perspective applicable knowledge* when participants adequately related concepts of the theoretical text to the sub-problems of the problem case that have not been regarded to as part of the core problem, but have been seen as necessary for a complete analysis of the problem case by an expert solution.

*Focused applicable knowledge as a co-construct* has proven to be a reliable measure (Cronbach's $\alpha = .77$). Although it depicts several independent sub-problems by definition, *multi-perspective applicable knowledge* has also proven to be reliable (Cronbach's $\alpha = .66$). The individual acquisition of *focused* and *multi-perspective applicable knowledge* was measured as being satisfactorily reliable with Cronbach's $\alpha = .66$, and respectively Cronbach's $\alpha = .55$.

For a better comparability with the outcomes regarding both the co-constructs and the individual acquisition of focused and multi-perspective applicable knowledge, all outcome measures will be reported as z-scores.

## 6.6    Control Variables

Control variables were assessed with the help of pen and paper tests and questionnaires that included multiple choice items and a baseline case. The data for the control variables were collected prior to the experiment except for computer-specific attitudes, which were measured after the collaborative phase in order to avoid priming effects. The individual control variables included:

(1) *Demographic data.* Several demographic variables, namely gender, age, and first language (and years speaking German as a foreign language) have been assessed with a questionnaire.

(2) *Prior knowledge.* Prior knowledge has been assessed with respect to applicable knowledge. Prior applicable knowledge has been assessed with a baseline case, comparable to the problem cases of the collaborative phase and the transfer case. For this reason, the 'coding system for a multi-level analysis of knowledge co-construction' has been applied to the pre-test case analyses. Raters were again blind with respect to experimental condition. 629 units of analysis have been segmented and categorized regarding applicable prior knowledge. With the students' analysis of the problem case, focused and multi-perspective applicable prior knowledge was measured. However, the 1st-year-students disposed of extremely little applicable prior knowledge. Thus, 90% of the students were unable to construct a relation that was attributable to focused applicable knowledge, and 76% of the students could not produce relations attributable to multi-perspective applica-

ble knowledge. Three coding units regarding focused as well as multi-perspective applicable knowledge were shown as the empirical maximum of individuals' prior knowledge. Due to this floor effect, possible differences regarding focused (Cronbach's $\alpha = .49$) and multi-perspective applicable knowledge (Cronbach's $\alpha = .33$) could not be reliably measured.

(3) *Learning strategies*. The learning strategies were measured with the help of Wild and Schiefele's (1994) scale prior to the experiment. Reliability was measured with Cronbach's Alpha ($\alpha = .64$).

(4) *Social anxiety*. Social anxiety was measured with the help of Rost and Schermer's (1997) scale prior to the experiment. Reliability was measured with Cronbach's Alpha ($\alpha = .92$).

(5) *Uncertainty orientation*. Uncertainty orientation has been assessed prior to the experiment as well, with Cronbach's $\alpha = .72$ (Dalbert, 1996).

(6) *Interest*. The interest towards the learning environment has been measured with a scale of some self-developed items in a questionnaire prior to the experiment based on Prenzel, Eitel, Holzbach, Schoenhein, and Schweiberer (1993; see also Krapp, 1999), e.g., "Please mark the statements that apply to you with an × (from 'does not apply' to 'does apply'): 'I am interested in getting to know new pedagogical theories and concepts'" (Cronbach's $\alpha = .74$).

(7) *Computer-specific attitudes*. In order to not prime participants with respect to learning with computers, participants were asked after the experiment to report their computer-specific attitudes regarding their belief of how computers may de-personalize society, based on a questionnaire by Richter et al. (2001); Cronbachs $\alpha = .77$.

## 6.7    Treatment Check

It has been checked if the treatments were realized by the participants in the intended way. Both SCOS- and ECOS-prompts should have
been answered according to the intention of the individual prompt. For instance, the SCOS-prompt "WE HAVE NOT REACHED CONSENSUS
CONCERNING THESE ASPECTS:" should have been followed by difference of opinions between the learning partners. In other words, the learning
partners were supposed to engage in conflict-oriented consensus building. If
the learners engaged in other social modes, e.g., if the response to this
prompt signaled quick consensus building, the prompt has been coded as
'not answered in the intended way' regardless of possible reasons for not
responding in the intended way, e.g., lack of knowledge. Therefore, the
treatment check consisted of the assessment of responses to the prompt that
diverged from the intention of the prompt. Additionally to unintended responses, missing responses to prompts were counted and entered the treatment check. The results of the treatment checks are calculated in relation to
the number of prompts of the individual conditions.

Additionally, SCOS guided learners through the individual discussion boards of the problem cases and pre-structured the number of the messages that the participants should contribute. This number of messages was
the same for all participants (eight messages in total). Therefore, the number
of messages and the heterogeneity of the number of messages will be analyzed as additional treatment checks of the SCOS. As an indicator for heterogeneity of the number of messages within the groups, dissimilarity scores
based on standard deviations of the number of messages will be analyzed
(cf. Cooke, Salas, Cannon-Bowers, & Stout, 2000; Fischer & Mandl, 2001;
Weinberger, Stegmann, & Fischer, 2007).

## 6.8      Statistic Procedures

All main and interaction effects will be tested with univariate ANOVA for statistical significance. Significant findings of an univariate ANOVA will be reported with regard to their effect sizes of explained variance ($\eta^2$). Post-hoc group comparisons will be calculated with unpaired t-tests. For correlation of the single process dimensions with the outcomes of collaborative knowledge construction, multiple linear regressions will be conducted. Effect sizes will be reported with reference to the explained variance with the adjusted $R^2$-value. Relations between outcomes as co-construct and as individual acquisition will be calculated with Pearson's bivariate correlation.

As the individual learners influence each other in the learning groups, all tests will be calculated based on groups of three, as unit of analysis. In order to explore relations between processes and individually acquired knowledge an exception to this group level procedure will be made because aggregation bias can be expected (Sellin, 1990). A hierarchical linear model may not be applied due to high requirements of this statistic procedure, e.g., large number of aggregated units (Ditton, 1998; Mok, 1996). Relations between processes and knowledge as co-construct will not be calculated, as variables of both dimensions feed from the same data in the collaborative phase.

As the total amount of analysis segments of the data sources are not identical, processes and learning outcomes will be graphically presented as z-scores to improve comparability. The z-transformations have been calculated over the whole sample. An α-level of .05 was used for all statistical tests, except differences between learning prerequisites. In order to warrant, that the groups did not differ regarding learning prerequisites, an α-level of .10 was used instead.

## 6.9 Case Studies

In addition to quantitative analyses, case studies with detailed descriptions and interpretations of four discourses will be presented – one for each experimental condition. The case studies aim to illustrate the actual collaboration of learners on the web-based discussion boards. In this way, the specific social and cognitive processes of collaborative knowledge construction will be qualitatively visualized in the specific discourse contexts in which they emerge (Chi, 1997; Yin, 2003). The case studies may explain more closely why and how the various social modes and epistemic activities are more or less essential components of collaborative knowledge construction. The case studies serve further to more generally evaluate discourse structures regarding transactivity and frequency of epistemic versus non-epistemic activity. In order to illustrate discourse structures, the case studies provide an overview of complete discourses with the help of a graphical coding analysis with coarser granularity of the units of analysis than in the quantitative studies.

### 6.9.1 Procedure of the case studies

The case studies follow qualitative analysis procedures (cf. Chi, 1997; Yin, 2003). First, the data of the conversational activities will be presented. This includes the graphical overview of the completed discourse, as was available onscreen to the learners. Furthermore, the individual messages will be listed in the temporal order they were posted (which does not necessarily correspond to the order of the graphical overview). Second, the discourses are analyzed in detail with respect to social and cognitive processes of collaborative knowledge construction. Finally, the case studies will be interpreted, and with the help of a graphical coding analysis, social and cognitive processes of collaborative knowledge construction will be allocated in the respective discourse structures.

### Data of the conversational activities

The case studies are based on complete discourses of learning groups about one problem case. The data of the conversational activities include the code names of the participants authoring the messages, the time the messages were posted, the position of the messages on the web-based discussion board, and the written words of the messages themselves. Figures of the onscreen overview of the completed discourses, as they were available to the participants, will be presented. These figures include information about the title, author, time, and position of the messages on the web-based discussion board (see figure 6.3.3d). This information will additionally be provided with each message text in order to make the messages identifiable with reference to the overview figure.

*Authors of the messages.* The learners will be referred to with their *code names* on the web-based discussion board, namely Ahorn, Birke, and Pinie. These code names were provided to all participants of the study in order to warrant anonymity. Of course, the participants of each case study were not identical, but were simply given the same code names in each of the separate experimental collaborative learning sessions.

*Time the messages were posted.* Information about the time the individual messages were posted is provided in the overview figures. In order to help identify the messages in the overview figures, the exact time the messages were sent off will also be reported when the complete texts of the messages are presented.

*Position of the messages on the web-based discussion board.* Messages on the web-based discussion board can take various positions, depending on whether they have been posted as a message that initiates a discussion thread or as a reply to a message. Initiating messages are located on the left of the screen on top of prior initiating messages. In this way, the initiating message that was posted last will be located at the top left. When a message is posted as a reply, it is placed below the message it is replying to, indented to the right, and graphically connected with a dotted line. This graphical connection builds a discussion thread. A discussion thread may consist of several, increasingly indented messages when replies are being responded to.

*Text of the message.* The actual text of the message starts with the information that is provided in the overview figure (title, author, position in the discussion thread, and the time it was posted). This information will be printed in bold letters. The actual message text may contain text that is automatically quoted from prior messages. This text is marked with an '>' and printed in shaded gray in order to improve identification of the new words of a message. Some messages in the script conditions also contain prompts. These prompts are written in capital letters.

### Social and cognitive processes in the case studies

After presenting the data of the conversational activities, the data will be interpreted in reference to the theoretical framework of the study. First, the social processes of the discourse will be analyzed. This analysis includes an evaluation of the discourse with respect to transactivity. Furthermore, the social modes of co-construction will be identified and analyzed with respect to their individual function within the discourse structure. Second, the cognitive processes that show in discourse will be analyzed. The discourses will be evaluated with respect to the frequency of epistemic versus non-epistemic activities. Subsequently, the various kinds of epistemic activities will be analyzed in detail. Beyond the identification of the various epistemic activities, the discourses will be evaluated with respect to the quality of learners' analyses. In this way, typical comprehension failures of students regarding Weiner's (1985) attribution theory may be discovered.

### Graphical Coding Analysis

The case studies will be illustrated with the help of a graphical coding analysis (cf. Keefer et al., 2000). This graphical coding analysis aims to provide an overview of the social and cognitive processes in the separate case studies. For this reason, the units of the graphical coding analysis are the separate messages.

The messages are represented with various shapes depending on the main *epistemic activity* they represent. Construction of problem space is indicated by hexagons, whereas construction of conceptual space is represented by diamonds. Squares indicate construction of relations between conceptual and problem space. Oval shapes represent non-epistemic activity (see figure 6.9.1a).

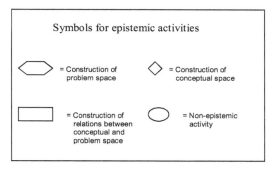

*Figure 6.9.1a*: Symbols for epistemic activities

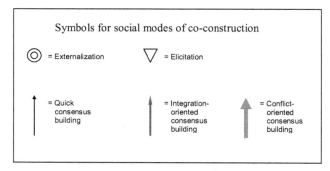

*Figure 6.9.1b*: Symbols for social modes of co-construction

*Social modes of co-construction* will be represented either within the message symbols or as connections between the message symbols (see figure 6.9.1b). The social mode of externalization will be indicated with a ring inside the respective message symbol. Elicitation is indicated with an upside down triangle within the message that contains the elicitation. The messages will be connected with arrows of various lines indicating the respective social modes of consensus building. The arrows start from the message that is showing the respective social mode and points to the contribution it is referring to. A weak arrow indicates quick consensus building, a medium-sized arrow indicates integration-oriented consensus building, and a thick arrow indicates conflict-oriented consensus building.

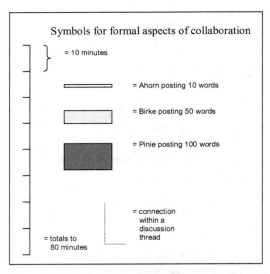

*Figure 6.9.1c*: Symbols for formal aspects of collaboration

Beyond social and cognitive processes, some more information is displayed in the graphical coding analysis. Apart from the social modes of consensus building, messages may be connected because they have been posted as replies to initiating messages. These connections are indicated by dotted lines and mirror discussion threads as represented by the web-based discussion board. The columns of the graphical coding analysis will display what position the separate messages take within the discussion threads. The number of new words indicating the length of messages, is displayed by the length of the message symbols. One millimeter of length of the message symbols represents ten new words that a message contains. The width of a message symbol does not hold information. Furthermore, the messages are posted at different times. This time is displayed vertically, starting from the first message on top of the graphic to the last one on the bottom. One millimeter represents one minute. Messages that lack new words will be presented with ×. Finally, the colors of the message symbols represent the individual participants Ahorn, Birke, and Pinie (see figure 6.9.1c).

*Interpretation of the case studies*

Finally, the case studies will be interpreted with respect to the specific research questions. In this way, discourse structures will be identified in reference to the graphical coding analysis. Furthermore, the quality of the discourse will be discussed in reference to comprehension failures of the learners, and in reference to the functionality of the social and cognitive process phenomena. Based on the theoretical framework of the study, the significance of specific social and cognitive processes with respect to collaborative knowledge construction will be discussed.

## 6.9.2 Selection of the discourses for the case studies

Discourses on one of three problem cases for each experimental condition were selected for the case studies. The case studies aim to illustrate the quantitative results. In order to select case studies that well represented the script effects, discourses of learning groups were randomly chosen from the two middle quartiles of summed up focused and multiperspective individually acquired knowledge. From the three discourses that resulted from the three problem cases that each group was given, the discourse, which most clearly showed specific social and cognitive processes, was selected as the case study. The selected discourses were either about the problem case "Asia" or about the problem case "Math." The problem case "Asia" is about Asian students having more beneficial attribution patterns than Western students (see figure 6.3.3d). The problem case "Math" is about a student who is subject to various attributions regarding his in-class-failures of mathematics (see section 6.3.2 and figure 6.3.3c). The selected discourse without script (DWS) is mainly characterized by quick consensus building and non-epistemic activities. The discourse facilitated with the social script (SCOS) is mainly characterized by elicitation, conflict-oriented consensus building, and construction of relations between conceptual and problem space. The discourse that was facilitated with the epistemic script (ECOS) is characterized by externalization and construction of relations between conceptual and problem space. Finally, the discourse facilitated with both scripts (ESCOS) is characterized by quick consensus building and non-epistemic activities.

# 7     Results of the Empirical Study

Results of the study will be reported on grounds of all 96 participants. The experimental groups will be abbreviated with DWS for discourse without structure induced by cooperation scripts (= control group), ECOS for epistemic cooperation script, SCOS for social cooperation script, and ESCOS for the combination of both epistemic and social cooperation script. First, results of the treatment check will be reported. Second, the research questions will be quantitatively analyzed. Finally, one case study will illustrate each experimental condition.

## 7.1     Treatment Check

On average, about 60 % of the prompts were responded to in the intended sense. No substantial differences with respect to the usage of the prompts could be found between the three experimental groups that were facilitated by prompts ($\chi^2_{(2)} = 2.48$, *n. s.*). This analysis is based on a comparison of the groups of three with any form of cooperation script ($n = 24$). With respect to the number of messages, the main effect of the SCOS can be considered to be substantial and large ($F_{(1,28)} = 16.05$; $p < .05$; $\eta^2 = .36$). Furthermore, no effect of the ECOS ($F_{(1,28)} = 2.89$; *n. s.*) and no interaction effect ($F_{(1,28)} = 2.99$; *n. s.*) can be found. The participants of the DWS-group (control) and the ECOS-learners authored more messages than the SCOS-learners, which wrote about eight messages (= 24 messages within a learning group of three), with or without additional ECOS, which was intended (see table 7.1). The smaller deviations from the suggested 24 messages in the SCOS-conditions can be explained by handling mistakes or messages that were written in addition to script suggestions. With respect to heterogeneity of the number of messages, a large main effect of the SCOS can be found ($F_{(1,28)}$ = 19.45; $p < .05$; $\eta^2 = .41$). Beyond that, neither a main effect of the ECOS ($F_{(1,28)}$ = 0.03; $n.\ s.$) nor an interaction effect ($F_{(1,28)}$ = 0.29; $n.\ s.$) can be found. While DWS- and ECOS-discourse without SCOS-support appear to be heterogeneous within the learning groups, with respect to the amount of messages sent, the number of messages were equally distributed over all group members in the SCOS-conditions (see table 7.1). The effect sizes point to the fact that the SCOS clearly determined the number of messages being sent. Therefore it can be said that in general, the treatments have been used in the intended way.

*Table 7.1:*    Number of messages in the individual experimental settings

| | **DWS** | **ECOS** | **SCOS** | **ESCOS** |
| | *M*  (*SD*) | *M*  (*SD*) | *M*  (*SD*) | *M*  (*SD*) |
| --- | --- | --- | --- | --- |
| Number of messages | 49.13 (18.72) | 35.00 (13.58) | 25.63 (2.07) | 25.50 (1.93) |
| Heterogeneity of number of messages | 4.69 (2.84) | 4.10 (3.67) | 0.55 (0.68) | 0.86 (0.66) |

## 7.2    Effects of Scripts on Processes of Collaborative Knowledge Construction

In this section, effects of the treatments on social and cognitive processes of collaborative knowledge construction in the text-based CSCL environment will be reported. The data have been collected on the web-based discussion boards of one of the three problem cases in the collaborative phase which lasted 80 minutes for all four experimental conditions.

### 7.2.1 Effects of scripts on social processes

First, the main effects of the social script, the epistemic script, and the interaction of both scripts regarding the social modes of co-construction, externalization, elicitation, quick consensus building, integration-oriented consensus building, and conflict-oriented consensus building will be presented. These social modes have been distributed unevenly over the discourse and between the individual experimental groups (see table 7.2.1).

*Table 7.2.1:* The means of the five social modes of co-construction in the groups of three in the individual experimental conditions (percentages may not add up to 100 % due to truncation).

| | DWS | ECOS | SCOS | ESCOS |
|---|---|---|---|---|
| | *M* (*SD*) | *M* (*SD*) | *M* (*SD*) | *M* (*SD*) |
| | % | % | % | % |
| **Externalization** | 4.41 (4.02) | 11.71 (5.74) | 4.75 (7.33) | 4.93 (7.80) |
| | 31 % | 57 % | 25 % | 26 % |
| **Elicitation** | 2.00 (2.37) | 1.00 (1.67) | 2.25 (2.77) | 1.26 (1.53) |
| | 14 % | 5 % | 12 % | 7 % |
| **Quick consensus building** | 4.04 (4.22) | 3.04 (4.62) | 5.12 (5.36) | 5.44 (7.98) |
| | 28 % | 15 % | 27 % | 29 % |
| **Integration-oriented consensus building** | 0.59 (1.53) | 0.75 (3.27) | 0.54 (1.38) | 0.85 (2.66) |
| | 4 % | 4 % | 3 % | 4 % |
| **Conflict-oriented consensus building** | 3.30 (3.48) | 4.17 (4.83) | 5.96 (5.50) | 6.44 (5.62) |
| | 23 % | 20 % | 32 % | 34 % |

*Externalization*

First of all, the effects of the two factors and their combination are presented with regard to *externalization*. With respect to externalization, an effect of the ECOS ($F_{(1,28)} = 24.10$; $p < .05$; $\eta^2 = .46$), an effect of the SCOS ($F_{(1,28)} = 11.95$; $p < .05$; $\eta^2 = .30$), and an interaction effect of both treatments ($F_{(1,28)} = 18.13$; $p < .05$; $\eta^2 = .39$) can be seen. All of these effects are large. The ECOS has a large positive effect on externalization. Learners in the ECOS-group showed three times as much externalization than any other group. More than half of their discourse activity can be described as thinking aloud in front of the group. Additional social structure clearly reduces the effect, however, to a low level achieved by the SCOS alone as well (see figure 7.2.1a).

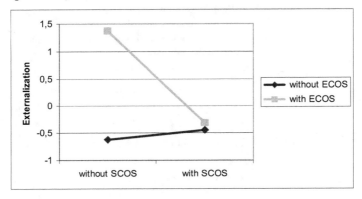

*Figure 7.2.1a*: Effects of the scripts on externalization.

### Elicitation

With respect to elicitation, a main effect of the ECOS could be found ($F_{(1,28)} = 3.46$; $p < .05$; $\eta^2 = .11$), but no main effect of the SCOS ($F_{(1,28)} = 0.14$; *n. s.*) or an interaction effect ($F_{(1,28)} = 0.14$; *n. s.*) can be found (see figure 7.3.1b). Thus, the SCOS cannot foster elicitation beyond the scope of elicitation as phenomenon in unscripted, open discourse. The ECOS, however, substantially reduces elicitation. This is a medium-sized effect. The descriptive data (see table 7.2.1) suggests that this was also the case for the ESCOS-condition, because ESCOS-learners produce equally little elicitation.

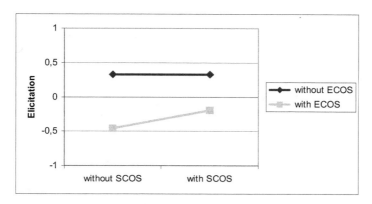

*Figure 7.2.1b*: Effects of the scripts on elicitation.

### Quick consensus building

With respect to quick consensus building, a SCOS-effect ($F_{(1,28)} = 3.32$; $p < .05$; $\eta^2 = .11$), but no ECOS ($F_{(1,28)} = 0.03$; *n. s.*) or interaction effect ($F_{(1,28)} = 0.98$; *n. s.*) can be found.

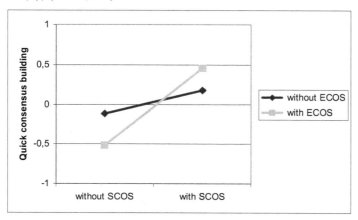

*Figure 7.2.1c*: Effects of the scripts on quick consensus building.

The SCOS-effect is medium-sized and positive, suggesting that the SCOS facilitates quick consensus building. Even though the diagram suggests an interaction effect (see figure 7.2.1c), the significance testing does not confirm an interaction effect. This might be explained by the rather high standard deviations (see table 7.2.1). The descriptive data further suggests that the ECOS-learners generally engaged little in any form of consensus building. This becomes particularly clear when the values are being proportionally considered.

### Integration-oriented consensus building

With respect to integration-oriented consensus building, no effects of the SCOS ($F_{(1,28)} = 0.00$; *n. s.*), the ECOS ($F_{(1,28)} = 0.16$; *n. s.*), or an interaction effect ($F_{(1,28)} = 0.02$; *n. s.*) can be found. Integration-oriented consensus building – which turned out to be a relatively rare and not a homogeneously distributed phenomenon – is not affected by any treatment or combination of treatments (see table 7.2.1). The z-scores for integration-oriented consensus building are $M = -.04$ ($SD = .65$) for DWS, $M = .06$ ($SD = 1.49$) for ECOS, $M = -.11$ ($SD = .55$) for SCOS, and $M = .09$ ($SD = 1.20$) for ESCOS.

### Conflict-oriented consensus building

With respect to conflict-oriented consensus building, a substantial main effect of the SCOS ($F_{(1,28)} = 4.10$; $p < .05$; $\eta^2 = .13$), but no ECOS-effect ($F_{(1,28)} = 0.15$; *n. s.*), or an interaction effect ($F_{(1,28)} = 0.09$; *n. s.*) can be found (see figure 7.2.1d). The SCOS-effect is medium and indicates that the SCOS substantially facilitates conflict-oriented consensus building.

Considering the descriptive, proportional data, about one third of the utterances that the learners who were supported with SCOS produced were conflict-oriented. In contrast, only about every fifth utterance of learners who were not supported with SCOS was conflict-oriented (see table 7.2.1).

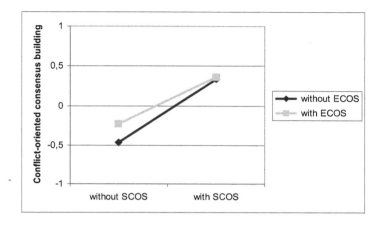

*Figure 7.2.1d*: Effects of the scripts on conflict-oriented consensus building.

## 7.2.2      Effects of scripts on cognitive processes

The effects of the two factors, ECOS and SCOS, and their interaction are presented below with regard to (non-)epistemic activities.

The epistemic activities were clearly unevenly distributed over the discourse in favor of the epistemic activity of constructing relations between conceptual and problem space (see table 7.2.2). The main epistemic activity of learners (90% and above) was the construction of relations between conceptual and problem space in the distinct experimental conditions.

Furthermore, the descriptive statistics suggest a difference between the experimental group without cooperation scripts vs. with cooperation scripts regarding all epistemic and non-epistemic activities, as was hypothesized. The mean of epistemic activities for the DWS-groups is $M = 10.08$ ($SD = 6.84$) in contrast to $M = 16.67$ ($SD = 15.17$) for groups supported with any form of script. The mean of non-epistemic activities for the DWS-groups is $M = 4.54$ ($SD = 3.72$) in contrast to $M = 2.94$ ($SD = 2.68$) for groups supported with any form of script. A post-hoc analysis proves this difference to be substantial with respect to epistemic activities ($t_{(85.68)}$ = -2.92; $p < .05$) and regarding non-epistemic activities ($t_{(94)}$ = 2.28; $p < .05$).

*Table 7.2.2*:  Epistemic activities in contrast to non-epistemic activities and the respective epistemic activities.

| | DWS | ECOS | SCOS | ESCOS |
|---|---|---|---|---|
| | *M* (*SD*) | *M* (*SD*) | *M* (*SD*) | *M* (*SD*) |
| | % | % | % | % |
| **Epistemic vs. non-epistemic activities** | | | | |
| **Non-epistemic activities** | 4.22 (3.64) | 3.17 (3.94) | 2.92 (1.74) | 2.48 (1.93) |
| | 29 % | 15 % | 16 % | 13 % |
| **Epistemic activities in total** | 10.11 (6.81) | 17.50 (10.38) | 15.71 (15.98) | 16.44 (17.63) |
| | 71 % | 85 % | 84 % | 87 % |
| **The respective epistemic activities*** | | | | |
| Construction of problem space | 0.44 (0.75) | 1.67 (1.90) | 0.46 (0.83) | 0.74 (2.01) |
| | 4 % | 10 % | 3 % | 5 % |
| Construction of conceptual space | 0.33 (0.68) | 0.13 (0.34) | 0.08 (0.28) | 0.07 (0.27) |
| | 3 % | 1 % | 0.5 % | 0.5 % |
| Construction of relations between conceptual and problem space | 9.33 (6.67) | 15.71 (9.90) | 15.17 (15.59) | 15.63 (16.62) |
| | 92 % | 90 % | 97 % | 95 % |

*Percentages refer to the epistemic activities in total and may not add up to 100 % due to truncation.

### Construction of problem space

With respect to the construction of problem space, a substantial main effect of the ECOS ($F_{(1,28)}$ = 5.41; $p < .05$; $\eta^2$ = .16), but no SCOS-effect ($F_{(1,28)}$ = 1.10; *n. s.*), or an interaction effect of both factors ($F_{(1,28)}$ = 1.64; *n. s.*) can be found (see figure 7.2.2a).

The negative results of the significance testing of interaction may also be led back to the relatively high standard deviations of the single values (see table 7.2.2). The ECOS substantially facilitated the construction of problem space. This ECOS-effect proves to be large.

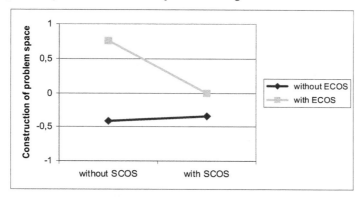

*Figure 7.2.2a*: Effects of the scripts on the construction of problem space.

### Construction of conceptual space

Regarding the construction of conceptual space, a substantial main effect of the SCOS ($F_{(1,28)} = 4.07$; $p < .05$; $\eta^2 = .13$), but no ECOS-effect ($F_{(1,28)} = 2.29$; *n. s.*), or an interaction effect ($F_{(1,28)} = 2.29$; *n. s.*) can be found (see figure 7.2.2b).

The interaction is probably not significant due to extremely high standard deviations of the typically rare phenomenon of construction of conceptual space (see table 7.2.2). The SCOS significantly reduced the construction of conceptual space. This SCOS-effect is medium-sized.

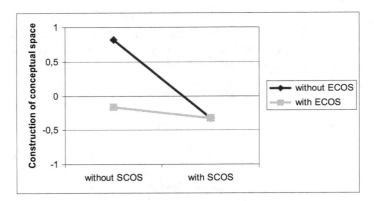

*Figure 7.2.2b*: Effects of the scripts on the construction of conceptual space.

*Construction of relations between conceptual and problem space*

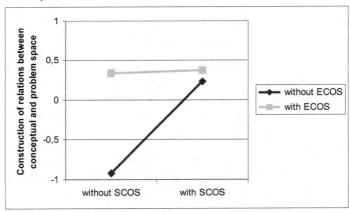

*Figure 7.2.2c*: Effects of the scripts on the construction of relations between conceptual and problem space.

With respect to the construction of relations between conceptual and problem space an effect of the ECOS ($F_{(1,28)} = 4.99$; $p < .05$; $\eta^2 = .15$), an effect of the SCOS ($F_{(1,28)}$ = 3.58; $p$ < .05; $\eta^2$ = .11), and an interaction effect ($F_{(1,28)}$ = 3.20; $p$ < .05; $\eta^2$ = .10) can be found (see figure 7.2.2c). The effects can be considered to be large, regarding the ECOS-effect and medium, regarding the SCOS and the interaction effects. Apparently, any script treatment facilitated the construction of relations between conceptual and problem space. The interaction effect shows, however, that this facilitation appears to have a peak limit – the effects of both treatments do not add up.

## 7.3 Effects of Scripts on Outcomes of Collaborative Knowledge Construction

In this section, the outcomes of knowledge co-construction with respect to focused and multi-perspective applicable knowledge as co-construct and as individual acquisition will be reported. First of all, the raw data regarding the outcomes will be presented (see table 7.3). The separate forms of knowledge can not be compared based on the raw values. Therefore, data illustrating the effects of scripts will be presented in z-scores.

*Table 7.3*: Raw values of learning outcomes regarding (focused and multi-perspective) knowledge as co-construct and (focused and multi-perspective) individually acquired knowledge.

| | **DWS** | **ECOS** | **SCOS** | **ESCOS** |
|---|---|---|---|---|
| | *M* (*SD*) | *M* (*SD*) | *M* (*SD*) | *M* (*SD*) |
| **Knowledge as co-construct** | | | | |
| Focused | 1.21 (1.69) | 2.96 (3.18) | 2.04 (2.99) | 2.29 (3.31) |
| Multi-perspective | 2.50 (2.17) | 2.75 (2.98) | 2.50 (3.13) | 3.17 (3.70) |
| **Individually acquired knowledge** | | | | |
| Focused | 4.10 (2.47) | 2.36 (2.22) | 3.93 (1.79) | 3.43 (2.48) |
| Multi-perspective | 2.29 (1.57) | 1.28 (1.45) | 4.22 (3.08) | 1.76 (2.02) |

## 7.3.1    Effects of scripts on knowledge as co-construct

In the following paragraph, the effects of both factors and their combination, with regard to focused and multi-perspective applicable knowledge as co-construct in the collaborative phase, will be examined.

First of all, the results show that the learners clearly applied more knowledge than in the individual pre-tests. In the pre-test, 90% of the participants were unable to construct a relation that was attributable to focused applicable knowledge, and 76% of the participants could not produce relations attributable to multi-perspective applicable knowledge (cf. section 6.6). During the collaborative phase, however, only 37% of the participants did not construct any relation that shows focused applicable knowledge, and 27% did not show any multi-perspective applicable knowledge. The empirical maximum for individuals regarding focused applicable knowledge were 13 coding units, the empirical maximum for individuals regarding multi-perspective applicable knowledge were 12 coding units. The minimum was 0 units for both focused and multi-perspective knowledge as co-construct.

### *Focused applicable knowledge as co-construct*

With respect to focused applicable knowledge as co-construct, a substantial effect of the ECOS ($F_{(1,28)} = 6.44$; $p < .05$; $\eta^2 = .19$), but no effect of the SCOS ($F_{(1,28)} = 0.63$; *n. s.*), nor an interaction effect ($F_{(1,28)} = 0.63$; *n. s.*) can be found (see figure 7.3.1).

This ECOS-effect is large. The epistemic script ECOS has facilitated focused applicable knowledge as co-construct. That means, that learners, who were supported with ECOS, could better analyze the cases collaboratively with respect to the elementary case information.

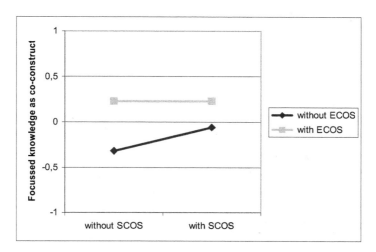

*Figure 7.3.1*: Effects of the scripts on focused applicable knowledge as a co-construct.

### *Multi-perspective applicable knowledge as co-construct*

With respect to multi-perspective applicable knowledge as a co-construct, no effects of the SCOS ($F_{(1,28)}$ = 0.00; *n. s.*), the ECOS ($F_{(1,28)}$ = 0.16; *n. s.*), or an interaction effect ($F_{(1,28)}$ = 0.02; *n. s.*) can be found. That means, that learners were not facilitated by any script or combination of scripts to analyze the problem cases together based on multi-perspective applicable knowledge. The z-scores for multi-perspective applicable knowledge were *M* = -.07 (*SD* = .73) for DWS, *M* = .01 (*SD* = 1.01) for ECOS, *M* = -.07 (*SD* = 1.06) for SCOS, and *M* = .15 (*SD* = 1.25) for ESCOS.

## 7.3.2    Effects of scripts on individually acquired knowledge

In the following section, the effects of both scripts and their combination with respect to the individual acquisition of applicable knowledge will be presented with z-scores for better comparability with the applicable knowledge as a co-construct. Applicable knowledge will be presented with respect to focused and multi-perspective applicable knowledge.

First of all, the participants clearly showed more applicable knowledge than in the pre-test and on a similar level as in the collaborative phase (cf. sections 6.6 and 7.3.1). Only 29% of the participants were not able to show focused applicable knowledge and 49% of the participants could not show multi-perspective applicable knowledge. The range for both focused and multi-perspective applicable knowledge was from min = 0 to max 10 coding units for individual learners. The participants only showed slightly more applicable knowledge in the collaborative phase. However, in the collaborative phase, learners were also still provided with the theoretical text in contrast to the individual post-test.

*Individually acquired focused applicable knowledge*

With regard to focused applicable knowledge, no main effects of the ECOS ($F_{(1,28)} = 1.59$; *n. s.*), and the SCOS ($F_{(1,28)} = 0.03$; *n. s.*), respectively, could be found. However, the interaction effect of both factors proves significant and large ($F_{(1,28)} = 5.60$; $p < .05$; $\eta^2 = .17$; see figure 7.3.2a). With the combination of both scripts, a level of individually acquired focused applicable knowledge can be reached that is comparable to the level of the DWS-control-group.

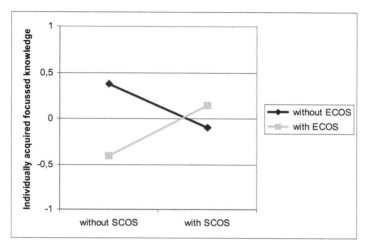

*Figure 7.3.2a*: Effects of the scripts on individually acquired focused applicable knowledge.

### Individually acquired multi-perspective applicable knowledge

Regarding multi-perspective applicable knowledge, an effect of the
ECOS can be found ($F_{(1,28)} = 6.89$; $p < .05$; $\eta^2 = .20$), as well as an effect of
the SCOS ($F_{(1,28)} = 3.56$; $p < .05$; $\eta^2 = .11$), but no interaction effect is observable ($F_{(1,28)} = 1.32$; *n. s.*). The SCOS has a positive medium-sized effect
on the individual acquisition of multi-perspective applicable knowledge. A
large effect is produced by the ECOS that impedes the acquisition of multi-
perspective applicable knowledge (see figure 7.3.2b).

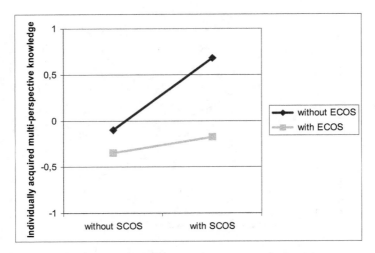

*Figure 7.3.2b*: Effects of the scripts on individually acquired multi-
perspective applicable knowledge.

## 7.4 Relations between Processes and Outcomes of Collaborative Knowledge Construction

In this section, the relations between variables examined in the collaborative phase (social processes, cognitive processes, and knowledge as
co-construct) and individual acquisition of applicable knowledge will be
reported. Multiple linear regressions and bivariate correlations will be calculated. The multiple linear regressions will be calculated on the basis of non-
overlapping categories that cover 100% of each of the two process dimensions. Furthermore, relations between knowledge as co-construct and individually acquired knowledge will be reported. The relations will be explored
on individual levels.

### 7.4.1 Relations between social processes and outcomes of collaborative knowledge construction

In this section, the social modes of co-construction and their relation to individual acquisition of applicable knowledge will be examined. Multiple linear regressions will be calculated including the five social modes of co-construction, namely externalization, elicitation, quick consensus building, integration-oriented consensus building, and conflict-oriented consensus building with regard to individual acquisition of applicable knowledge.

Regarding individually acquired applicable knowledge an individual level analysis shows that conflict-oriented consensus building relates positively with focused applicable knowledge, but explains little variance ($F_{(1, 94)} = 4.55$; $p < .05$; $R^2_{adj.} = .04$). With regard to individually acquired multiperspective applicable knowledge, a substantial positive relation to elicitation of an equally small effect size can be found ($F_{(1, 94)} = 4.27$; $p < .05$; $R^2_{adj.} = .03$).

### 7.4.2 Relations between cognitive processes and outcomes of collaborative knowledge construction

In this section, epistemic activities will be portrayed regarding their relation to individual acquisition of applicable knowledge. Multiple linear regressions are calculated, including construction of problem space, construction of conceptual space, construction of relations between conceptual and problem space, and non-epistemic activities with respect to individual acquisition of applicable knowledge.

An individual level analysis shows that a model consisting of positive relations of construction of conceptual space and constructions of relations between conceptual and problem space strongly relates to focused applicable knowledge ($F_{(2, 93)} = 64.06$; $p < .05$; $R^2_{adj.} = .57$). The β-score of construction of conceptual space in this model is $\beta = .14$ and the β-score of construction of relations between conceptual and problem space is $\beta = .76$.

With regard to individually acquired multi-perspective applicable knowledge, a substantial and strong relation to construction of relations between conceptual and problem space can be found ($F_{(1, 94)}$ = 52.80; $p < .05$; $R^2_{adj.}$ = .35). The effect sizes of these substantial relations are large. Epistemic activities, characterized by individual learners referring to conceptual space, relate to individually acquired focused applicable knowledge. Relations between conceptual and problem space relate to the individual acquisition of multi-perspective applicable knowledge.

### 7.4.3    Relations between knowledge as co-construct and individual acquisition of knowledge

In table 7.4.3, the bivariate correlation of focused and multi-perspective applicable knowledge as co-construct with individually acquired focused and multi-perspective applicable knowledge is presented. The members of groups of learners, who applied more multi-perspective knowledge collaboratively, have often individually acquired more focused applicable knowledge. No other substantial relations were found.

*Table 7.4.3:*    Relations of knowledge as co-construct with individually acquired knowledge.

|  | Focused knowledge as co-construct | Multi-perspective knowledge as co-construct |
| --- | --- | --- |
| Focused applicable knowledge as individual acquisition | -.09 | .37* |
| Multi-perspective applicable knowledge as individual acquisition | -.18 | -.06 |

* $p < .05$ tested two-sided

## 7.5 Case Studies

In the following sections, one case study for each experimental condition will be presented. First of all, a discourse without cooperation script (DWS) will be reported. Subsequently, a discourse that was facilitated with the social script (SCOS) and a discourse that was facilitated with the epistemic script (ECOS) will be presented. Finally, a discourse that was facilitated with both social and epistemic scripts will be presented (ESCOS). Within the separate case studies, the data of the conversational activities will be presented first. The presentation of the discourse data aims to closely resemble the appearance of the discourses on the web-based discussion boards. Mistakes in spelling of the participants, for example, will not be corrected. The discourses will be presented in complete form regarding one of the problem cases that the participants were expected to discuss and analyze. This includes the overview of the web-based discussion board and the full text of the individual messages as they were also available to the participants of the study. Subsequently, the social and cognitive processes observable in the discourse will be analyzed and the data of each case study will be interpreted. In order to visualize the social and cognitive processes, a graphical coding analysis will be applied to each case study.

### 7.5.1 DWS – discourse without cooperation script

First of all, a discourse of a learning group that was not facilitated with a cooperation script will be analyzed. The problem case requests students to explain and discuss why Asian students produce better results in mathematical tests than American and European students do on grounds of attribution theory. Ahorn in this DWS-group was a 25 year old female in her first semester at the university. Birke was a 20 year old female student in her first semester. Pinie was a 24 year old female Czech student, with German as her second language. She had 5 years German speaking experience. Pinie was in her second semester.

*Data of the conversational activities*

As figure 7.5.1a shows, there is one discussion thread with two more messages aiming to initiate a new discussion thread, which were not responded to. This discourse was conducted on the afternoon of the 23$^{rd}$ of January 2001.

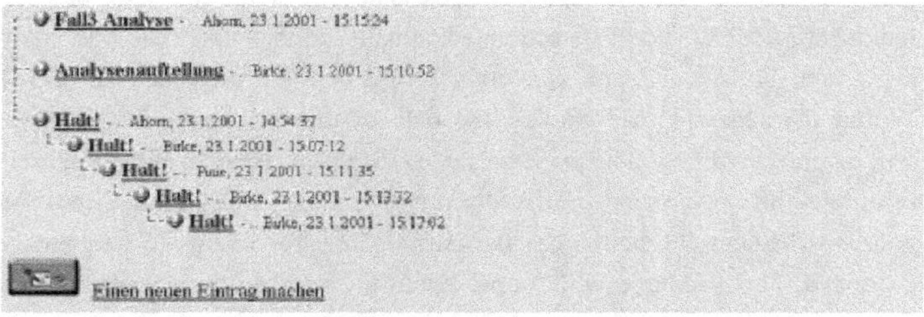

*Figure 7.5.1a*: Overview of a discourse without cooperation script

The first message of this discussion was produced by Ahorn at 14:54:37 titled "Halt!". In this message, Ahorn attempted to coordinate the learning group and suggested to further analyze another problem case.

| 1st message titled "Halt!" by Ahorn posted at 14:54:37 |
| --- |
| Ich bin mir nicht sicher, ob wir den Fall Nachhilfelehrer schon ausdiskutiert haben! Gehen wir nochmal zurück! |

This message was responded to by Birke about 13 minutes later. Notably, Birke did not refer to the initiating message, but started to analyze the case.

| 2nd message in the „Halt!" discussion thread posted by Birke at 15:07:12 |
| --- |
| > Ich bin mir nicht sicher, ob wir den Fall<br>> Nachhilfelehrer schon ausdiskutiert haben! Gehen wir nochmal<br>> zurück!<br>Vielleicht haben die Asiaten generell eine andere Einstellung und werden schon dahingehend erzogen nicht an ihrer eigenen Begabung zu zweifeln |

Only a few minutes later, Birke posted a new message, which was not responded to. In this message, Birke also attempted to coordinate task activities regarding another problem case.

> **3rd message initiating a new discussion thread titled „Auf-gabenaufteilung" posted by Birke at 15:10:52**
>
> Haben nur noch wenig Zeit gehe zurück zu Mathe
> und schreibe Analyse

With the fourth message, the third learning partner, Pinie, entered the discussion. Pinie suggested an alternative explanation for the problem case.

> **4th message in total and 3rd message in the "Halt!" discussion thread posted by Pinie at 15:11:35**
>
> >> Ich bin mir nicht sicher, ob wir den Fall
> >> Nachhilfelehrer schon ausdiskutiert haben! Gehen wir nochmal
> >> zurück!
> > Vielleicht haben die Asiaten generell eine andere
> > Einstellung und werden schon dahingehend erzogen nicht an
> > ihrer eigenen Begabung zu zweifeln
> Es ist auch möglich, dass die Lehrer durch
> Evaluation von den Schülern sich bemühen das Interesse und
> Motavation hoch zu erhalten. - Reattributionstraining.Pinie

Birke suggested that both aspects were valid and declared to briefly address these aspects.

> **5th message in total and 4th message in the "Halt!" discussion thread posted by Birke at 15:13:32**
>
> >>> Ich bin mir nicht sicher, ob wir den Fall
> >>> Nachhilfelehrer schon ausdiskutiert haben! Gehen wir nochmal
> >>> zurück!
> >> Vielleicht haben die Asiaten generell eine andere
> >> Einstellung und werden schon dahingehend erzogen nicht an
> >> ihrer eigenen Begabung zu zweifeln
> > Es ist auch möglich, dass die Lehrer durch
> > Evaluation von den Schülern sich bemühen das Interesse und
> > Motavation hoch zu erhalten. - Reattributionstraining.Pinie
> Ok alles beides dannschreibe ich halt schnellwas
> hier zu

In the meantime, Ahorn, who posted the first message in this web-based discussion board, composed an analysis of her own. Again, Ahorn's message was not responded to, even though Ahorn was referring to some aspects of the partners' analyses.

> **6th message initiating a new discussion thread titled "Fall3 Ana-lyse" posted by Ahorn at 15:15:24**
>
> Aufgrund Ihrer Erziehung haben die Asiatischen
> Kinder ein positiveres Attributionsverhalten. Das heißt
> sie erklären Mißerfolge nicht mit mangelnder eigener
> Begabung sondern mit mangelnder Anstrengung und die
> Erfolge entsprechend.Ich würde auch sagen, daß das auf
> die Erziehung zurückzuführen ist!

Finally, Birke translated her proposal to juxtapose both Pinie's, and her own case analyses into action.

> **7th message in total and 5th message in the "Halt!" discussion thread posted by Birke at 15:17:02**
>
> >>>> Ich bin mir nicht sicher, ob wir den Fall
> >>>> Nachhilfelehrer schon ausdiskutiert haben! Gehen wir nochmal
> >>>> zurück!
> >>> Vielleicht haben die Asiaten generell eine andere
> >>> Einstellung und werden schon dahingehend erzogen nicht an
> >>> ihrer eigenen Begabung zu zweifeln
> >> Es ist auch möglich, dass die Lehrer durch
> >> Evaluation von den Schülern sich bemühen das Interesse und
> >> Motavation hoch zu erhlaten. - Reattributionstraining Pinie
> > Ok alles beides dannschreibe ich halt schnellwas
> > hier zu
> Wahrscheinlich werden Asiaten schon früh auf ein
> gutes Selbstbewußtsein getrimmt und besitzen dadurch
> ein gutes Selbstbewußtsein und eine "stabile"
> Überzeugung von sich selbst. Außerdem lassen es die Lehrer
> gar nicht zu, dass die Schüler an sich selbst
> zweifeln.(Reattributionstraining) und wohl auch die Eltern unter'stütze

### Social processes

Several social processes can be identified in this discourse example. The participants interacted with each other in different ways. With the first message, Ahorn aimed to *elicit* some reaction from her learning partners, asking them to return to another problem case. Birke's reply to this message was not actually responding to this elicitation. Instead, she *externalized* her perspective upon the case. Subsequently, Birke posted a message externalizing her next steps, which also concerned another problem case. Again, this message was not responded to. The next message was the only message to contain a direct reference to a prior message: Pinie posed an alternative explanation, and thus, engaged in *conflict-oriented consensus building*. The participants did not resolve this conflict, but juxtaposed the two perspectives and *quickly built* a *consensus*. Pinie did not comment further on Birke's attempt to juxtapose Pinie's analysis to her own. Ahorn's second message, which *externalized* another analysis of the case, was also being ignored. In the last message, Birke composed an analysis referencing both Pinie's prior analysis and her own. These perspectives were not linked to an integrated perspective upon the problem case. Instead, Birke signaled quick consensus when referring to her learning partners and indicated that both aspects were correct.

To what extent did these unscripted learning partners engage in transactive discourse? There are several indications that this discourse can be referred to as *hardly transactive*. Several messages appear to be completely ignored. In particular, several messages that contain *elicitation* were not followed by any response. Interestingly enough, this includes all messages from one member of the learning group (Ahorn). This student appeared to be somewhat closed out of the group processes even though she tried to relate to the partners' contributions. Furthermore, Birke's coordinating move was not responded to. When referring to contributions of learning partners, Birke signaled quick consensus at all times. In this way, the one more transactive, conflict-oriented discourse move that Pinie made did not become a starting point for the learners to discuss multiple perspectives upon the case. Instead, the analyses were not linked, but accumulated. In this way, the discourse did not exceed each student's individual performance. The overall structure of this discourse was determined by quick consensus building and unanswered elicitation. The learners did not build on each others' contributions. On the contrary, learners seemed to be disconnected from each other. One of the learners was being ignored; another one only contributed one message. The learners did not systematically build on each others' contributions, but rather appeared to engage in random, ad-hoc interactions fulfilling minimal requirements regarding consensus building.

### Cognitive processes

With respect to cognitive processes, the extent to which learners engaged in *epistemic activities* can be analyzed. In the discourse above, learners frequently needed to coordinate their activities. The coordination concerning the other two cases as can be seen in the $1^{st}$ and in the $3^{rd}$ message, as well as in the present case (see $5^{th}$ message). These messages can be regarded as *non-epistemic activities*. Thus, the learners only engaged in epistemic activities to analyze the problem case in four of the seven messages.

The learners only engaged in one specific kind of epistemic activity. The learners neither portrayed the respective principles of attribution theory they wanted to apply, nor did they attempt to discuss aspects of the presented case problem. Instead, immediately after coordinating themselves, learners *constructed relations between conceptual and problem space*. Even though Birke and Pinie considered two different perspectives, these approaches to solve the problem case only marginally relate to attribution theory. Birke argued that Asians "generally have a different attitude" and Pinie suggested that "Asian teachers aim to foster motivation." These learners quickly converged towards these possible explications of the case. The quality of their analyses can be characterized as being punctual and incomplete, because they do not refer well to Weiner's (1985) attribution theory. Furthermore, the attribution theory terminology was used inadequately. Birke suggested that Asians "dispose of a 'stable' self-concept." However, stability in Weiner's (1985) attribution theory refers to characteristics of ascribed causes of success or failure instead of more or less stable personality traits. The learners did not continuously develop better conceptions of the problem case, but rather aimed to satisfy minimal requirements to present some kind of scarcely elaborated analysis. One of the participants, Ahorn, produced a more adequate analysis of the case. Ahorn recognized that Asians may show different, more beneficial attribution behavior (instead of determined personality traits). However, Ahorn's message was not responded to and her considerations did not enter Birke's final analysis. On top of that, the learners poorly managed the time allotted to analyze the case. The discussion took place in the last 20 minutes of the overall 80 minutes allowed for the collaborative phase. They quickly worked on the task at a superficial level and mainly aimed to coordinate their activities in the separate virtual spaces.

### Interpretation of the discourse data

The discourse as a whole appears to be unbalanced in regard to several aspects (see figure 7.5.1b). Regarding both social and cognitive demands of the learning task, this discourse represents *satisficing*. Contributions were unevenly spread, contributions and learners were ignored, and consensus was quickly built. The quality of the analyses is low. Learners did not explore problem space or conceptual space at all. The learners offered instead separate shots at how the case could be analyzed. The separate, incomplete, and inaccurate analyses were not well connected. The learners did not build shared conceptions of the problem. Instead, learners showed suboptimal management of their time and the virtual spaces they moved in. Furthermore, they appeared to be occupied, to some extent, with coordinating group activities.

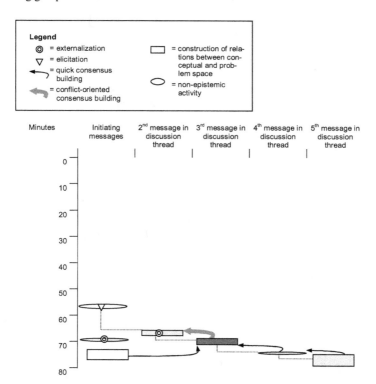

*Figure 7.5.1b*: Graphical coding analysis of a discourse without scripts.

Additionally, some comprehension failures emerge regarding the use of attribution theory terminology. This may indicate that two out of the three learners may not have understood the central idea of attribution theory. The central idea of attribution theory states that personal success or failure may be explained by stable/variable and internal/external causes, and that these explanations or *attributions* may be more or less beneficial for further learning. Instead, the participants consider the causes as being actual personality traits or *attributes* that may be directly more or less beneficial for learning.

## 7.5.2      SCOS – discourse with social cooperation script

In this section, a discourse that was facilitated with the social cooperation script is discussed. Again, the case is about the advantage of Asian over Western students in mathematics that learners should discuss based on attribution theory. Ahorn in the present SCOS-case-study was 24 years old, female, and in her first semester. Birke was a 25 year old female in her first semester. Pinie was 24 years old, female, and in her second semester. The first language of all participants was German.

### *Data of the conversational activities*

The discourse took place on the evening of the 17th of January 2001. At a first glance, the SCOS-learners kept exactly to the given structure as the treatment check has suggested (see figure 7.5.2a).

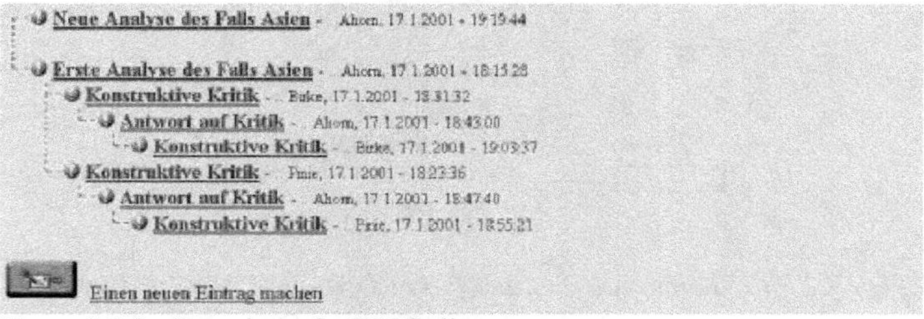

*Figure 7.5.2a*: Overview of a SCOS-discourse

A case analyst wrote a first analysis of the case, the two constructive critics answered this first analysis, the case analyst answered both critiques, the critics again responded to their respective answer, and finally, the case analyst composed a new analysis of the case. Furthermore, the learners did not modify the message titles which were set by the SCOS. Therefore, the first message is titled "Erste Analyse des Falls Asien" and the further messages are called "Konstruktive Kritik," "Antwort auf Kritik," and finally "Neue Analyse des Falls Asien." The initial analysis was posted at 18:15:28 by the case analyst Ahorn.

---

**1st message titled "Erste Analyse des Falls Asien" posted by Ahorn at 18:15:28**

Die asiatischen Eltern haben anscheinend die Misserfolge eher auf die Aufgabenschwierigkeit zurückgeführt. Im Fallbeispiel steht leider nicht, wie die amerikanischen und europäischen Eltern und Schüler sich Erfolge und Misserfolge erklärt haben. Aber offensichtlich hatten sie ungünstigere Attributionsmuster: Sie haben vielleicht Misserfolge auf mangelnde Begabung zurückgeführt oder Erfolge auf den Zufall. Aber wie gesagt, das steht hier nicht drin. Man sollte wohl ein Reattributionstraining mit den amerikanischen und europäischen Eltern und Schülern machen.

---

The next message was posted 8 minutes later as was assured by the SCOS by participant "Pinie." Pinie responded to all prompts of the constructive critic by giving some instructions to Ahorn about how to analyze the case. These instructions are either representing a general task strategy like "Focus on the task!" or "Give explanations!" or can be described as vague content-specific hints towards "Cultural context of the case." Pinie did not pick up specific aspects of Ahorn's analysis.

---

**2nd message titled "Konstruktive Kritik" posted by Pinie at 18:23:36**

> Die asiatischen Eltern haben anscheinend die
> Misserfolge eher auf die Aufgabenschwierigkeit zurückgeführt.
> Im Fallbeispiel steht leider nicht, wie die
> amerikanischen und europäischen Eltern und Schüler sich Erfolge
> und Misserfolge erklärt haben. Aber offensichtlich
> hatten sie ungünstigere Attributionsmuster: Sie haben
> vielleicht Misserfolge auf mangelnde Begabung zurückgeführt
> oder Erfolge auf den Zufall. Aber wie gesagt, das
> steht hier nicht drin. Man sollte wohl ein
> Reattributionstraining mit den amerikanischen und europäischen Eltern
> und Schülern machen.

FOLGENDE PUNKTE SIND MIR NOCH UNKLAR:

Ausführung ist sehr unausführlich und nicht themenflächendeckend genug.
PUNKTE BEI DENEN WIR NOCH NICHT EINER MEINUNG SIND:
Es liegt meiner Meinung nach eher am Umfeld und dessen Struktur und kulturellem Aufbau und

---

MEINE VORSCHLÄGE FÜR EINE ÄNDERUNG DER ANALYSE:
Geh mehr auf die eigentliche Frage ein und
analysiere die Leistungsunterschiede-Problematik. Gib mehr
Beispiele für die besseren Attributionsmerkmale.

The third message was again posted 8 minutes later by "Birke." Again all critic prompts were considered by "Birke." Birke, at first, wondered why Ahorn qualified her statements. In this way, she criticized formal aspects of Ahorn's initial analysis. Subsequently, Birke pointed out further explanations for the case that complement Ahorn's analysis on a concrete level.

**3rd message titled "Konstruktive Kritik" posted by Birke at 18:31:32**

> Die asiatischen Eltern haben anscheinend die
> Misserfolge eher auf die Aufgabenschwierigkeit zurückgeführt.
> Im Fallbeispiel steht leider nicht, wie die
> amerikanischen und europäischen Eltern und Schüler sich Erfolge
> und Misserfolge erklärt haben. Aber offensichtlich
> hatten sie ungünstigere Attributionsmuster. Sie haben
> vielleicht Misserfolge auf mangelnde Begabung zurückgeführt
> oder Erfolge auf den Zufall. Aber wie gesagt, das
> steht hier nicht drin. Man sollte wohl ein
> Reattributionstraining mit den amerikanischen und europäischen Eltern
> und Schülern machen.

FOLGENDE PUNKTE SIND MIR NOCH UNKLAR:
Warum zögerst Du so mit Deiner Antwort? Ich
glaube schon, dass Du teilweise recht hast.

PUNKTE BEI DENEN WIR NOCH NICHT EINER MEINUNG
SIND:
Ich glaube, dass es noch weitere
Erklärungsmöglichkeiten gibt. Ich denke, dass die Kinder ihren
Misserfolg
weniger auf die Aufgabenschwierigkeit zurückführen (die
können sie ja nicht verändern) als vielmehr darauf, dass
sie sich zu wenig angestrengt haben.

MEINE VORSCHLÄGE FÜR EINE ÄNDERUNG DER ANALYSE:

Ich glaube, wenn Du schreibst, dass die Kinder
die Ursache in der mangelnden Anstrengung sehen,
lässt sich auch deren höhere Motivation erklären.

As suggested by the social cooperation script, the fourth and fifth messages were composed by the case analyst, Ahorn, and are replies to the constructive critiques. First of all, Ahorn replied to Birke's critique using all prompts of the case analyst that respond to the critic prompts. In her case analyst reply, Ahorn first of all explained why she qualified her statements. Subsequently, Ahorn actually continued to discuss the case based on Birke's suggestions and refined her initial analysis. Furthermore, Ahorn aimed to construct conceptual space regarding a specific dimension of the attribution theory, eliciting Birke's help.

---

**4th message titled "Antwort auf Kritik" posted by Ahorn at 18:43:00**

>> Die polnischen Eltern haben anscheinend die
>> Misserfolge aber auf die Aufgabenschwierigkeit zurückgeführt.
>> Im Fußballspiel sieht leider nicht, wie die
>> amerikanischen und europäischen Eltern und Schüler sich Erfolge
>> und Misserfolge erklärt haben. Aber offensichtlich
>> hatten sie ungünstigere Attributionsmuster. Sie haben
> vielleicht Misserfolge auf mangelnde Begabung zurückgeführt
>> oder Erfolge auf den Zufall. Aber wie gesagt, das
>> steht hier nicht drin. Man sollte wohl ein
>> Reattributionstraining mit den amerikanischen und europäischen Eltern
>> und Schülern machen.
> FOLGENDE PUNKTE SIND MIR NOCH UNKLAR
> Warum zögerst Du so mit Deiner Antwort? Ich
> glaube schon, dass Du teilweise recht hast.
> PUNKTE BEI DENEN WIR NOCH NICHT EINER MEINUNG
> SIND.
> Ich glaube, dass es noch weitere
> Erklärungsmöglichkeiten gibt. Ich denke, dass die Kinder Ihren Misserfolg
> weniger auf die Aufgabenschwierigkeit zurückführen (das
> können sie ja nicht verändern) als vielmehr darauf, dass
> sie sich zu wenig angestrengt haben.
> MEINE VORSCHLÄGE FÜR EINE ÄNDERUNG DER ANALYSE
> Ich glaube, wenn Du schreibst, dass die Kinder
> die Ursache in der mangelnden Anstrengung sehen,
> lässt sich auch deren höhere Motivation erklären.

ZU DEN UNKLARHEITEN:
Ich habe gezögert, weil im Fall steht, dass sich
die günstigeren Attributionsmuster "im Hinblick auf
die Dimensionen der Stabilität" erklären lassen. Da
dachte ich, dass nur internal stabile und external
stabile Attributionen darunter fallen.

ZU UNSEREN MEINUNGSVERSCHIEDENHEITEN:
Wenn variable Attributionen da auch gemeint sind,
denke ich, dass Du recht hast - das wär dann internal
variable Attribution.

ZU DEN ÄNDERUNGSVORSCHLÄGEN:
Und dann wäre die Erklärung für die höhere
Motivation auch richtig.

Wie siehst denn Du das? Ist mit "Dimensionen der
Stabilität" gemeint, dass es sich nur um stabile
Attributionen handelt oder auch um variable? Andererseits
hätten sie sich dann den Spruch auch sparen können.

---

Subsequently, Ahorn replied to Pinie's critique, basically using all prompts of the case analyst that respond to the critic prompts, but implying that the last prompt was responded to earlier. Ahorn appeared to have spent less time answering Pinie's critique. Basically, Ahorn defended herself against Pinie's critique pointing out that there is only little case information.

She also elicited Pinie to contribute more concrete suggestions regarding her critique.

---

**5th message titled "Antwort auf Kritik" posted by Ahorn at 18:47:40**

>> Die asiatischen Eltern haben anscheinend die
>> Misserfolge eher auf die Aufgabenschwierigkeit zurückgeführt.
>> Im Fallbeispiel steht leider nicht, wie die
>> amerikanischen und europäischen Eltern und Schüler sich Erfolge
>> und Misserfolge erklärt haben. Aber offensichtlich
>> hatten sie ungünstigere Attributionsmuster: Sie haben
>> vielleicht Misserfolge auf mangelnde Begabung zurückgeführt
>> oder Erfolge auf den Zufall. Aber wie gesagt, das
>> steht hier nicht drin. Man sollte wohl ein
>> Reattributionstraining mit den amerikanischen und europäischen Eltern
>> und Schülern machen.
> FOLGENDE PUNKTE SIND MIR NOCH UNKLAR:
> Ausführung ist sehr unausführlich und nicht
> themenflächendeckend genug.
> PUNKTE BEI DENEN WIR NOCH NICHT EINER MEINUNG
> SIND
> Es liegt meiner Meinung nach eher am Umfeld und
> dessen Struktur und kulturellem Aufbau und
> MEINE VORSCHLÄGE FÜR EINE ÄNDERUNG DER ANALYSE.
> Geh mehr auf die eigentliche Frage ein und
> analysiere die Leistungsunterschiede-Problematik. Gib mehr
> Beispiele für die besseren Attributionsmerkmale.
ZU DEN UNKLARHEITEN:

In dem Fallbeispiel steckt ja auch nicht
sonderlich viel Information. Nur dieses "vor allem
günstigere Attributionsmuster im Hinblick auf die
Dimensionen der Stabilität" und das hab ich erklärt. Meiner
Meinung nach, gibt es da nur die eine Erklärung. Schreib
mir doch Deine Vorschläge
ZU UNSEREN MEINUNGSVERSCHIEDENHEITEN:
Es geht doch um die Attributionsmuster. Und da
ist nur eine Information gegeben (so).
ZU DEN ÄNDERUNGSVORSCHLÄGEN:

so.

---

Eight minutes later, Pinie posted another constructive critique as suggested by the script. Pinie basically responded to Ahorn's elicitation for more concrete suggestions.

---

**6th message titled "Konstruktive Kritik" posted by Pinie at 18:55:21**

>>> Die asiatischen Eltern haben anscheinend die
>>> Misserfolge eher auf die Aufgabenschwierigkeit zurückgeführt.
>>> Im Fallbeispiel steht leider nicht, wie die
>>> amerikanischen und europäischen Eltern und Schüler sich Erfolge
>>> und Misserfolge erklärt haben. Aber offensichtlich
>>> hatten sie ungünstigere Attributionsmuster: Sie haben
>>> vielleicht Misserfolge auf mangelnde Begabung zurückgeführt
>>> oder Erfolge auf den Zufall. Aber wie gesagt, das
>>> steht hier nicht drin. Man sollte wohl ein
>>> Reattributionstraining mit den amerikanischen und europäischen Eltern
>>> und Schülern machen.

---

>> FOLGENDE PUNKTE SIND MIR NOCH UNKLAR
>> Ausführung ist sehr unausführlich und nicht
>> thematflächendeckend genug
>>PUNKTE BEI DENEN WIR NOCH NICHT EINER MEINUNG
>> SIND:
>> Es liegt meiner Meinung nach eher im Umfeld und
>>dessen Struktur und kulturellem Aufbau und
>> MEINE VORSCHLÄGE FÜR EINE ÄNDERUNG DER ANALYSE
>> Geh mehr auf die eigentliche Frage ein und
>> analysiere die Leistungsunterschiede-Problematik. Gib mehr
>> Beispiele für die besseren Attributionsmerkmale.
> ZU DEN UNKLARHEITEN:
> In dem Fallbeispiel steckt ja auch noch
> sonderlich viel Information. Nur dieses vor allem
> günstigere Attributionsmuster im Hinblick auf die
> Dimensionen der Stabilität und das hat ich erklärt. Meiner
> Meinung nach gibt es da nur die eine Erklärung. Schreib
> mir doch Deine Vorschläge
> ZU UNSEREN MEINUNGSVERSCHIEDENHEITEN:
> Es geht doch um die Attributionsmuster. Und da
> ist nur eine Information gegeben (so).
> ZU DEN ÄNDERUNGSVORSCHLÄGEN:
> so.

FOLGENDE PUNKTE SIND MIR NOCH UNKLAR:
Die Stabilität hat sowohl zeitlich stabile als
auch variable Ursachenmöglichkeiten. Geh auf beide
Gesichtspunkte ein.

PUNKTE BEI DENEN WIR NOCH NICHT EINER MEINUNG
SIND:
Alles was Du gesagt hast war richtig, nur versuch
Deine Analyse nicht so speziell zu halten. Mathematik
ist in Asien einer der wichtigsten Gesichtspunkte im
Schulwesen, berücksichtige auch das.

MEINE VORSCHLÄGE FÜR EINE ÄNDERUNG DER ANALYSE:
Die Kinder sehen die Ursache evtl. in der
mangelden Anstrengung. Deshalb ist die Motivation in Ihren
Fällen durchaus zu steigern.

Another eight minutes later, Birke posted another constructive critique as suggested by the script. Birke responded to Ahorn's elicitation and indicated further that Ahorn and Birke had come to some kind of agreement on how to understand the theory in order to analyze the case.

**7th message titled "Konstruktive Kritik" posted by Birke at 19:03:37**

>>> Die asiatischen Eltern haben anscheinend die
>>> Misserfolge eher auf die Aufgabenschwierigkeit zurückgeführt.
>>> Im Fallbeispiel steht leider nicht, wie die
>>> amerikanischen und europäischen Eltern und Schüler sich Erfolge
>>> und Misserfolge erklärt haben. Aber offensichtlich
>>> hatten sie ungünstigere Attributionsmuster. Sie haben
>>> vielleicht Misserfolge auf mangelnde Begabung zurückgeführt
>>> oder Erfolge auf den Zufall. Aber wie gesagt, das
>>> steht hier nicht drin. Man sollte wohl ein
>>> Reattributionstraining mit den amerikanischen und europäischen Eltern
>>> und Schülern machen.
>> FOLGENDE PUNKTE SIND MIR NOCH UNKLAR:
>> Warum zögerst Du so mit Deiner Antwort? Ich

>> glaube schon, dass Du teilweise recht hast.
>> PUNKTE BEI DENEN WIR NOCH NICHT EINER MEINUNG
>> SIND:
>> Ich glaube, dass es noch weitere
>> Erklärungsmöglichkeiten gibt. Ich denke, dass die Kinder ihren Misserfolg
>> weniger auf die Aufgabenschwierigkeit zurückführen (die
>> können sie ja nicht verändern) als vielmehr darauf, dass
>> sie sich zu wenig angestrengt haben.
>> MEINE VORSCHLÄGE FÜR EINE ÄNDERUNG DER ANALYSE:
>> Ich glaube, wenn Du schreibst, dass die Kinder
>> die Ursache in der mangelnden Anstrengung sehen,
>> lässt sich auch deren höhere Motivation erklären.
> ZU DEN UNKLARHEITEN:
> Ich habe gezögert, weil im Fall steht, dass sich
> die günstigeren Attributionsmuster "im Hinblick auf
> die Dimensionen der Stabilität" erklären lassen. Da
> dachte ich, dass nur internal stabile und external
> stabile Attributionen darunter fallen.
> ZU UNSEREN MEINUNGSVERSCHIEDENHEITEN:
> Wenn variable Attributionen da auch gemeint sind,
> denke ich, dass Du recht hast - das wär dann internal
> variable Attribution.
> ZU DEN ÄNDERUNGSVORSCHLÄGEN:
> Und dann wäre die Erklärung für die höhere
> Motivation auch richtig.
> Wie siehst denn Du das? Ist mit "Dimensionen der
> Stabilität" gemeint, dass es sich nur um stabile
> Attributionen handelt oder auch um variable? Andererseits
> hätten sie sich dann den Spruch auch sparen können.
FOLGENDE PUNKTE SIND MIR NOCH UNKLAR:

Ja, ich glaube, dass mit der Dimension der
Stabilität einfach gemeint ist, ob eine Ursache stabil oder
variabel ist.
PUNKTE BEI DENEN WIR NOCH NICHT EINER MEINUNG
SIND:

Ich glaube sie wollten damit nur sagen, dass wir
nicht auf die Lokalität eingehen sollten.
MEINE VORSCHLÄGE FÜR EINE ÄNDERUNG DER ANALYSE:

Ich habe eigentlich keine neuen Vorschläge, bis
auf dass Du Deine "vielleichts" eigentlich streichen
kannst. Ansonsten: siehe 1.Kritik.

Finally, Ahorn, the case analyst, composed a new analysis of the case 16 minutes later, as determined by the script. Ahorn referred to several aspects of her rather adequate first analysis, but also regarded the constructive critique that had been posed during discourse by Birke. This constructive critique also entered the final analysis.

8th message titled "Neue Analyse des Falls Asien" posted by
Ahorn at 19:19:44

Also: Wenn die asiatischen Eltern und Kinder vor
allem günstigere Attributionsmuster im Hinblick auf die
Dimensionen der Stabilität haben, sehen sie den Grund für
ihre Erfolge und Misserfolge in ihrer Anstrengung und
nutzen Fähigkeit und Begabung nicht als
Erklärungsmuster. So glauben sie bei jeder neuen Aufgabe, durch
große Anstrengung ein gutes Ergebnis bekommen zu
können. Der Zufall sollte bei ihnen nur als

Erklärungsmuster für Misserfolge dienen.

Sind ihre Attributionsmuster gerade im Hinblick
auf die Dimension der Stabilität ungünstiger als die
der asiatischen Eltern und Kinder, so begründen
amerikanische und europäische Eltern und Kinder ihre Leistungen
eher mit Begabung (was nie motivationsfördernd ist)
und bei Erfolgen mit der Aufgabenschwierigkeit. Der
Zufall könnte bei den Europäern und Amerikanern auch mal
als Erklärung für gute Ergebnisse herhalten müssen.

## Social processes

In the SCOS-condition, learners were expected to adopt specific roles and engage in specific social modes of co-construction. Have the learners engaged in the social modes that were suggested by the SCOS? First of all, Ahorn engaged in *externalization* in order to act out her role as case analyst. Subsequently, the constructive critics posted their messages. Both Birke and Pinie actually composed reviews that consisted of *elicitation* and *conflict-oriented consensus building*. Apparently, the learners adopted the social script suggestions well. Pinie criticized in general, that the first analysis of Ahorn needed to be elaborated. Subsequently, she hinted towards alternative aspects that Ahorn needed to consider. Finally, she explicitly advised Ahorn to focus on the actual question and give more examples. In her first message, Birke suggested that Ahorn should not qualify her statements, because she is right in some parts. Subsequently, Birke pointed out the aspects that have not yet been considered by Ahorn, thus following a typical 'yes-but' critique structure. Ahorn generally agreed with Birke's critique. It can be noted that Ahorn was about to *integrate* Birke's suggestions. Furthermore, Ahorn was inviting Birke to join in analyzing the case by asking a specific question concerning conceptual space and thus, also engaged in *elicitation*.

In contrast, Ahorn was defending herself against Pinie's critique. Again Ahorn was aiming to *elicit* more information from Pinie in spite of the fact that she had indicated that she did not agree with her critique. In this way, Ahorn engaged in *conflict-oriented consensus building*. Pinie was actually embarking on giving more detailed feedback and critically addressed the same aspects as Birke did. It is possible that Pinie had followed the parallel discussion thread of Ahorn and Birke, and adopted Birke's critique.

However, this was not indicated by Pinie. Pinie seriously engaged in the role of the critic, requesting that Ahorn also considers "that mathematics is an important aspect in the Asian school system." Notably, Pinie did not back up this statement with any proof. Pinie's critiques were apparently based on the authority of her externally induced social role as constructive critic rather than on the authority of proof warranting her claims. Considering that Pinie was studying only in her second semester of educational sciences and is the same age as her learning partners, she was displaying considerable certainty in her role as critic. For instance, her final statement was that "the motivation of the kids can, by all means, be increased." In the next message Birke also responded to Ahorn's elicitation. Furthermore, she flexibly used the second prompt (WE HAVE NOT REACHED CONSENSUS CONCERNING THESE ASPECTS:). This means, she did not point out another difference of opinion after the second prompt, but continued the discussion which had evolved within this 'difference-of-opinions-frame' constituted by the corresponding prompts. Finally, she again referred to her initial critique, which consisted of encouragement to take the collaboratively constructed analysis as granted, instead of qualifying her statements. Conclusively, Ahorn aimed to *build an integration-oriented consensus*. In the actual final analysis, Ahorn referred to various aspects which had been considered in the preceding discussions, instead of simply copying her initial analysis.

### Cognitive processes

In general, the learners engaged only in epistemic activities in contrast to non-epistemic activities. That means, the learners clearly aimed to jointly analyze the case. As the quantitative data suggests, the *construction of relations between conceptual and problem space* is the primary epistemic activity of the learners. The separate discussion threads of the two constructive critics display some differences, however, about how learners approach the case. Pinie vaguely referred to alternative explanations such as cultural context. Ahorn responded to this critique by pointing to missing case information, which can be regarded to as an attempt to *construct problem space*. In contrast, Birke directly complemented Ahorn's analysis. Ahorn integrated Birke's critique and subsequently aimed to *construct conceptual space* together with Birke. Ahorn and Birke appeared to construct a shared conception of the theory and its application in the concrete case. Clearly, Ahorn's conceptions regarding one dimension of attribution theory were enhanced. As a result of collaborative knowledge construction, Ahorn had begun to understand that the dimension 'stability' has two specifications, namely, being variable or stable. As a consequence, Ahorn was enabled to formulate a new analysis based on her initial, and basically adequate analysis, but also based on the knowledge that had been collaboratively constructed in discourse.

### Interpretation of the discourse data

First of all, the learners have reproduced the exact discourse structure provided by the SCOS, consisting of eight specific messages. Overall the discourse represents a highly coherent and transactive structure (see figure 7.5.2b). It can be seen that the SCOS-learners have not only referred to each other, but have also operated on and shared each others' conceptions to build better analyses of the case. The learners were able to productively apply the externally induced cooperation script. The scripted roles, however, were used flexibly. Ahorn did not take over and integrate any given form of critique. She accepted Birke's critique, but particularly defended her analysis against Pinie's critique. This may indicate that the externally induced SCOS needs to be represented in the learner's mind in a sensible, flexible way. Learners may need to understand the underlying principle of the script with respect to the learning goals in order to apply the script accordingly.

The social script apparently has facilitated learners to produce specific social modes of co-construction. There are indications, however, that learners may not be able to fulfill the requirements of the social roles being only supported with the respective prompts. The case study shows that learners may act as critics, but fail to understand problem space, conceptual space, and the relations between conceptual and problem space themselves. In this way, inadequate conceptions may enter the discourse and may be decisively represented due to the authority given by the social script. In the present case study, however, inadequate conceptions were not integrated. Instead, the case analyst was able to distinguish justified from misleading critique.

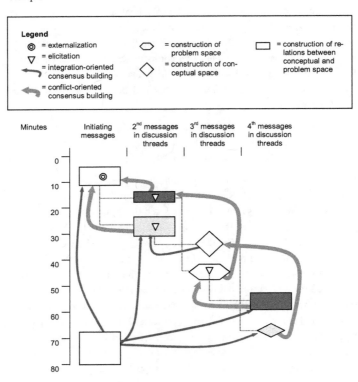

*Figure 7.5.2b*: Graphical coding analysis of a SCOS-discourse.

### 7.5.3 ECOS – discourse with epistemic cooperation script

In this section, a discourse that was facilitated with the epistemic co-operation script will be discussed. The problem case the students were re-quired to analyze and discuss is about a fictional student who is subject to various attributions regarding his in-class-failures of mathematics. See section 6.3.2 for a description of the case. Ahorn of the ECOS-case-study was a 19 year old, female in her first semester. Birke was 23 years old, female, and also in her first semester. Birke indicated that Bosnian was her first language and that she had spoken German for 7 years. Pinie was a 20 year old female student in her first semester.

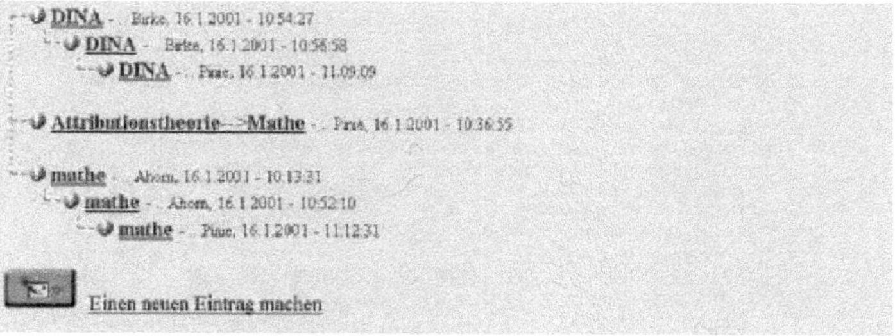

*Figure 7.5.3a*: Overview of an ECOS-discourse

### *Data of the conversational activities*

In the ECOS-discourse, which was conducted on the morning of the 16th of January 2001, there are three messages initiating discussion threads. At a first glance at the overview page (figure 7.5.3a), two of these messages appear to actually build discussion threads, i.e. they appear to be responded to. A closer inspection however, reveals that all of the replies lack new words. These empty postings may indicate that the learners have called up the messages without intention to answer, but to return to the overview page. Then, instead of clicking the back button, they pushed the response button and sent off an empty message in order to come back to the overview page. It is possible that these learners only remembered this more complicated way back to the overview page.

As the learners' responses have been left empty, there are actually only three messages. The first message was posted by Ahorn at 10:13:31 titled "mathe." In this message the first and the last prompt was not responded to. In this analysis several misconceptions were displayed regarding concepts of attribution theory and their application.

---

**1st message titled "mathe" posted by Ahorn at 10:13:31**

MIT DER ATTRIBUTIONSTHEORIE ERKLÄRBARE
FALLINFORMATIONEN:

FÜR DIESEN FALL RELEVANTE BEGRIFFE DER
ATTRIBUTIONSTHEORIE:
- GEHT DER ATTRIBUTION EIN ERFOLG ODER EIN
MISSERFOLG VORAUS?
es gehen mißerfolge voraus
- IST DIE LOKALITÄT DER ATTRIBUTION INTERNAL ODER
EXTERNAL?
die lokalität ist external, da die eltern ihm
einreden,daß er einfach nicht begabt.und internal, da er es
selber glaubt und das als ausrede nimmt
- WIRD AUF EINE STABILE ODER EINE VARIABLE
URSACHE ATTRIBUIERT?
die ursache ist variabel, weil er faul ist und
nur mit spickzettel auskommt
- ATTRIBUIERT DER BETROFFENE SELBST ODER WIRD VON
AUßEN FREMD ATTRIBUIERT?es wird fremd
attribuiert(lehrerin)

PROGNOSE UND PÄDAGOGISCHE KONSEQUENZEN AUS DER
PERSPEKTIVE DER ATTRIBUTIONSTHEORIE:
da der junge von den eltern immer bestätigt
bekommt, daß die ganze familie in mathe einfach nicht
begabt ist, wird er selbst sich nicht aufraffen und
anfangen zu lernen. die eltern schieben es einfach auf
seine begabung. diese meinung nimmt er natürlich an.man
muß ihm deutlich machen daß er begbt ist und eine
gute leistung in mathe erreichen kann, wenn er anfängt
zu lernen. er muß von seinen fähigkeiten überzeugt
und motiviert werden

NICHT MIT DER ATTRIBUTIONSTHEORIE ERKLÄRBARE
FALLINFORMATIONEN:

---

The next message was posted 23 minutes later. This message was initiating a new discussion thread. The quality of the analysis was extremely high and more or less resembled expert analyses of the case.

---

**2nd message initiating a new discussion thread titled "Attributionstheorie-->Mathe" posted by Pinie at 10:36:55**

---

MIT DER ATTRIBUTIONSTHEORIE ERKLÄRBARE
FALLINFORMATIONEN:
- Selbstattribution Michaels
- Meinung der Lehrerin
- Meinung der Eltern
- geringe Motivation Michaels

FÜR DIESEN FALL RELEVANTE BEGRIFFE DER
ATTRIBUTIONSTHEORIE:
- GEHT DER ATTRIBUTION EIN ERFOLG ODER EIN
MISSERFOLG VORAUS?
- Misserfolge
- IST DIE LOKALITÄT DER ATTRIBUTION INTERNAL ODER
EXTERNAL?
- internal
- WIRD AUF EINE STABILE ODER EINE VARIABLE
URSACHE ATTRIBUIERT?
- Michael und seine Eltern: stabile Ursache
(Begabung)
- Lehrerin: variable Ursache (Anstrengung)
- ATTRIBUIERT DER BETROFFENE SELBST ODER WIRD VON
AUßEN FREMD ATTRIBUIERT?
- Michael attribuiert selbst, nach seiner Meinung
ist die Ursache seine Begasbung (internal stabil)
- Seine Eltern geben die gleiche Ursache für
seinen Misserfolg an
- Die Lehrerin hält seine mangelnde Anstrengung
(internal variabel) für die Ursache

PROGNOSE UND PÄDAGOGISCHE KONSEQUENZEN AUS DER
PERSPEKTIVE DER ATTRIBUTIONSTHEORIE:
- Wenn Michael weiterhin seine Begabung als
Ursache für seinen Misserfolg hält, werden seine
Leistungen wohl kaum besser, da er denkt, dass er an dieser
Ursache sowieso nichts ändern kann.
- Wohingegen wenn er die Ursache des Misserfolgs
in der mangelnden Anstrengung erkennt, kann er indem
er sich mehr anstrengt seine Leistungen verbessern.

NICHT MIT DER ATTRIBUTIONSTHEORIE ERKLÄRBARE
FALLINFORMATIONEN:
- Dass er auf einem Schulfest trinkt.
- Dass Michael auf das nächste Schuljahr gespannt
ist.

The 3$^{rd}$ message was posted 16 minutes after the 2$^{nd}$ message and did not contain any words apart from the quoted message it was responding to. Apparently, the author herself read her analysis once more.

3rd message in total and 2nd message in the "mathe" thread lacking new words posted by Ahorn at 10:52:10

The 4$^{th}$ message was posted shortly afterwards, containing an analysis of the case by the third learning partner. There is no clear explanation why this discussion thread is titled "DINA." Most likely, the student wanted to indicate her actual name. The student marked the last prompt with "X" and thus did not respond to this prompt in the intended way. Birke basically appeared to have understood the underlying principles of attribution theory, but only focused on specific aspects of the problem case. The student wrote the complete message in capital letters. In order to distinguish the words of the participant from the prompts, all words of the student are presented in small letters.

---

**4th message in total initiating a new discussion thread titled "DINA" posted by Birke at 10:54:27**

MIT DER ATTRIBUTIONSTHEORIE ERKLÄRBARE
FALLINFORMATIONEN:
um etwas zu erziehlen muß man sich anstrengen
zeile1-3
EINE FREMDATTRIBUTION(DIE ELTERN)ALS AUSREDE
FÜR DIESEN FALL RELEVANTE BEGRIFFE DER
ATTRIBUTIONSTHEORIE:
- GEHT DER ATTRIBUTION EIN ERFOLG ODER EIN
MISSERFOLG VORAUS?
ein mißerfolg, da das kind denk seine eltern
"waren nicht gut"im mathe  und leben trozdem warum soll
ich mir dann so viele gedanken machen wie ich es
weiter machen soll
- IST DIE LOKALITÄT DER ATTRIBUTION INTERNAL ODER
EXTERNAL?
es ist internal,also es wird das alles als
begabun schwache angenommen zeile 3-8
- WIRD AUF EINE STABILE ODER EINE VARIABLE
URSACHE ATTRIBUIERT?
auf eine stabile die selben zeilen wie oben
- ATTRIBUIERT DER BETROFFENE SELBST ODER WIRD VON
AUßEN FREMD ATTRIBUIERT?
der betroffene wird selbst attribuiert aber auch
fremdattribuiert

PROGNOSE UND PÄDAGOGISCHE KONSEQUENZEN AUS DER
PERSPEKTIVE DER ATTRIBUTIONSTHEORIE:
da der betroffene von seine eltern
"unterschtutzt"ist bzw es werden keine gesprache entwickel wo die
eltern dem kind erklaren sollen wenn die nicht so gut
gewesen sind soll sich das kind etwas anstrengen um denen
auch spater vielleicht helfen zu konen und so das kind
motiviren kann es sein dass das kind nur so viel erziehlt
wie es notig ist um ins nächste jahr zu kommen.

NICHT MIT DER ATTRIBUTIONSTHEORIE ERKLÄRBARE
FALLINFORMATIONEN:
x

---

Any further postings were left empty. In the 5th message, the author called up her own message again. The last two empty messages were the third postings within the respective discussion threads.

---

**5th message in total and 2nd message in the "DINA" thread lacking new words posted by Birke at 10:56:58**

---

> 6th message in total and 3rd message in the "DINA" thread lacking new words posted by Pinie at 11:09:09

> 7th message in total and 3rd message in the "mathe" thread lacking new words posted by Pinie at 11:12:31

### Social processes

The portrayed ECOS-discourse is *minimally transactive*. Learners only engaged in *externalization*. Each learner had posted a more or less complete analysis of the case without referring to any message of a learning partner. In this regard, learners did not actually discuss with each other. If learners were able "to operate on each other's reasoning," this could not be traced in the individual analyses presented here. The students did not at all fulfill the requirement to build consensus.

### Cognitive processes

The individual analyses were *epistemic activities only*. As the first prompt suggested learners to construct problem space, this epistemic activity can be found in those analyses that have responded to this prompt. However, most reactions to this first prompt can also be described as an attempt by learners to *construct relations between conceptual and problem space*. The analyses of the learners lack construction of conceptual space. Instead, the learners aimed to immediately apply concepts to the case. The analyses differ with regard to their adequacy. Whereas Birke produced an analysis of the case which closely resembles an expert analysis, several *misconceptions* can be traced in Ahorn's analysis. Ahorn confused the locality concept (internal vs. external) with self-attribution vs. attribution of others. Furthermore, there are indications that Ahorn did not fully understand the conception of the cognitive attribution theory. She confers to attributes rather than attributions. Interestingly enough, this lack of understanding does not prevent her from making adequate prognoses and suggest proper pedagogical interventions. She appears to understand the educational implications of attribution theory without being able to abstract the principles and to construct an adequate mental model of the theory.

### Interpretation of the discourse data

In general, the activities of the learners in the ECOS-discourse took place parallel to each other with no mutual referencing whatsoever (see figure 7.5.3b). The qualities of the separate analyses vary greatly from high to low, with one learner continuously showing misconceptions, one learner referring to some aspects of the problem case, and one learner producing a complete and accurate analysis of the problem case.

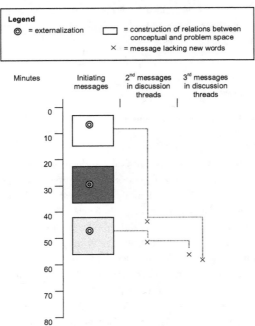

*Figure 7.5.3b*: Graphical coding analysis of an ECOS-discourse.

One may question whether these learners profited from each other. It is possible that they did not consider the approaches of their learning partners, continued to hold on to their point of view, and as a consequence, failed to acquire multiple perspectives. Furthermore, a typical comprehension failure can be identified.

One learner failed to understand that the attribution theory does not determine case characteristics, but what the individual believes to be the cause for personal success or failure. This appears to be a typical comprehension failure of learners dealing with attribution theory. Typically, these learners infer that the protagonist of the problem case is actually too 'lazy,' instead of analyzing, for instance, whether the protagonist himself believes that he is lazy and attributes school failure to this laziness. An example for this misconception of the attribution theory can be found in the present case study.

In conclusion, it can be said that not all members of a learning group may be able to adequately apply epistemic scripts. The epistemic script may support expertlike approaches of some learners, but cannot reduce misconceptions of those learners who have not understood the underlying principles of the attribution theory. Furthermore, the epistemic script may facilitate individual rather than collaborative knowledge construction. In the present case study, for instance, learners did not discuss their individual approaches at all. Therefore, epistemic scripts may not produce the specific outcomes that are associated with collaborative knowledge construction. Learners may not internalize multiple perspectives upon a subject matter, but adhere rigidly to more or less adequate initial approaches.

### 7.5.4    ESCOS – discourse with social and epistemic co-operation scripts

In this section a discourse that was facilitated with both epistemic cooperation script as well as social cooperation script is portrayed. The problem case that the students are supposed to analyze and discuss is again about the student who is subject to various more or less beneficial attributions regarding his failures in the subject mathematics. The case is completely quoted in section 6.2.2. Ahorn of the ESCOS-case-study was a 19 year old female in her first semester. Birke was a 29 year old female also in her first semester. Birke's first language was Hungarian. She had spoken German for 7 years. Pinie was a 21 year old female student in her first semester.

### *Data of the conversational activities*

As the social script component suggests, eight messages had been posted in the ESCOS-discourse which was conducted on the morning of the 12th of January 2001. The discourse structure resembles exactly the externally induced structure of one case analyst (Pinie) and two constructive critics (Ahorn and Birke). The learners did not modify the message titles which were set by the script (see figure 7.5.4a).

*Figure 7.5.4a*: Overview of an ESCOS-discourse

Pinie, the case analyst, started the discourse with the first analysis "Erste Analyse des Falls Mathe" at 10:44:05. In this first message, the epistemic component of the ESCOS-condition was realized. Pinie did not use the first prompt in the intended way to collect relevant case information that can be explained with attribution theory. Instead, she gave a concise and accurate analysis of the case. All following prompts were responded to in the intended way. The rest of the given analysis is somewhat redundant.

---

**1st message titled "Erste Analyse des Falls Asien" posted by Pinie at 10:44:05**

MIT DER ATTRIBUTIONSTHEORIE ERKLÄRBARE
FALLINFORMATIONEN:
michael und seine eltern machen die mangelnde
begabung, die wohl in der familie liegt, für die schlechten
matheleistungen verantwortlich - laut b. weiner also eine
internal stabile attribution.
seine lehrerin "beschuldigt" ihn der faulheit,
eine internal variable attribution.

FÜR DIESEN FALL RELEVANTE BEGRIFFE DER
ATTRIBUTIONSTHEORIE:
- GEHT DER ATTRIBUTION EIN ERFOLG ODER EIN
MISSERFOLG VORAUS?
misserfolg
- IST DIE LOKALITÄT DER ATTRIBUTION INTERNAL ODER
EXTERNAL?
internal, da die begabung beim schüler selbst
lokalisiert ist
- WIRD AUF EINE STABILE ODER EINE VARIABLE
URSACHE ATTRIBUIERT?
auf eine stabile
- ATTRIBUIERT DER BETROFFENE SELBST ODER WIRD VON
AUßEN FREMD ATTRIBUIERT?
beides ist der fall, denn michael selbst sagt,
daß mathe nicht "sein fach" ist (hier also die
selbstattribution). andererseits liegt auch fremdattribution vor, da
seine eltern ihn ebenfalls durch ursachenzuschreibungen
beeinflussen .
ausserdem attribuiert seine lehrerin von aussen,
indem sie sagt, m. wäre faul.

PROGNOSE UND PÄDAGOGISCHE KONSEQUENZEN AUS DER
PERSPEKTIVE DER ATTRIBUTIONSTHEORIE:
michael wird sich wohl kaum weiter anstrengen, da
er überzeugt ist, nichts an seinen leistungen im
fach mathe ändern zu können. die stabile variable ist
ja für ihn nicht veränderbar.

NICHT MIT DER ATTRIBUTIONSTHEORIE ERKLÄRBARE
FALLINFORMATIONEN:
michaels vater hatte ebenfalls untrricht bei der
gleichen lehrerin, dies ist aber meiner meinung nach keine
attribution.

---

The next message was posted eight minutes later at 10:51:45 as foreseen by the social component of ESCOS. The 'constructive critic,' Birke, sent only a message consisting of the prompts, but did not produce any new comments herself.

---

**2nd message titled "Konstruktive Kritik" lacking new words posted by Birke at 10:51:45**

---

The third message was posted by Ahorn at 10:58:35 which is a little earlier than suggested by the script. Ahorn did not use the full eight minutes of the critic's phase, but sent off the message prior to the end of the count-

down. As a response to the prompt "WE HAVE NOT REACHED CON-
SENSUS CONCERNING THESE ASPECTS:" Ahorn wrote "I completely
agree with you." Thus it can be said that this prompt had not been responded
to in the intended way. Concerning the desire for clarity and modification
proposals, Ahorn referred to the case information that Pinie had referred to
as not relevant for an analysis based on attribution theory. This case information points out that Ms Weber was a teacher for both the father and the
son. Ahorn asked Pinie to better explain her point regarding this case information.

> **3rd message titled "Konstruktive Kritik" posted by Ahorn at 10:58:35**
>
> \> MIT DER ATTRIBUTIONSTHEORIE ERKLÄRBARE
> \> FALLINFORMATIONEN:
> \> michael und seine eltern machen die mangelnde
> \> begabung, die wohl in der familie liegt, für die schlechten
> \> mathleistungen verantwortlich - laut b. weiner also eine
> \> internal stabile attribution.
> \> seine lehrerin "beschuldigt" ihn der faulheit,
> \> eine internal variable attribution.
> \> FÜR DIESEN FALL RELEVANTE BEGRIFFE DER
> \> ATTRIBUTIONSTHEORIE:
> \> - GEHT DER ATTRIBUTION EIN ERFOLG ODER EIN
> \> MISSERFOLG VORAUS?
> \> misserfolg
> \> - IST DIE LOKALITÄT DER ATTRIBUTION INTERNAL ODER
> \> EXTERNAL?
> \> internal, da die begabung beim schüler selbst
> \> lokalisiert ist
> \> - WIRD AUF EINE STABILE ODER EINE VARIABLE
> \> URSACHE ATTRIBUIERT?
> \> auf eine stabile
> \> - ATTRIBUIERT DER BETROFFENE SELBST ODER WIRD VON
> \> AUSSEN FREMD ATTRIBUIERT?
> \> beides ist der fall, denn michael selbst sagt,
> \> daß mathe nicht "sein fach" ist (hier also die
> \> selbstattribution). andererseits liegt auch fremdattribution vor, da
> \> seine eltern ihn ebenfalls durch ursachenzuschreibungen
> \> beeinflussen.
> \> ausserdem attribuiert seine lehrerin von aussen,
> \> indem sie sagt, m. wäre faul.
> \> PROGNOSE UND PÄDAGOGISCHE KONSEQUENZEN AUS DER
> \> PERSPEKTIVE DER ATTRIBUTIONSTHEORIE:
> \> michael wird sich wohl kaum weiter anstrengen, da
> \> er überzeugt ist, nichts an seinen leistungen in
> \> fach mathe ändern zu können. die stabile variable ist
> \> ja für ihn nicht veränderbar.
> \> NICHT MIT DER ATTRIBUTIONSTHEORIE ERKLÄRBARE
> \> FALLINFORMATIONEN:
> \> michaels vater hatte ebenfalls unterricht bei der
> \> gleichen lehrerin. dies ist aber meiner meinung nach keine
> \> attribution.
>
> FOLGENDE PUNKTE SIND MIR NOCH UNKLAR:
> Warum ist es keine Attribution des Vaters, wenn
> er bei der gleichen Lehrerin Unterricht hatte?

Chapter 7: Results of the Empirical Study

At 11:06:56 the case analyst Pinie responded to Ahorn's constructive critique. Pinie appreciated Ahorn's "complete agreement." Furthermore, Pinie began to discuss the case information that Ms Weber was a teacher to both the father and the son. Whilst discussing this case information, Pinie showed some inaccurate use of attribution theory terminology referring to 'attributes' instead of 'attributions.'

**4th message titled "Antwort auf Kritik" posted by Pinie at 11:06:56**

> Bin völlig deiner Meinung.
> MEINE VORSCHLÄGE FÜR EINE ÄNDERUNG DER ANALYSE:
> Letzten Punkt (besser) begründen! WARUM?
ZU DEN UNKLARHEITEN:
ich dachte,daß es irgendwie nicht mit den
attributionen zusmmenhaängt, es könnte aber auch sein, daß hier
das attribut 'zufall' vorliegt, was aber nicht
variavbel ist, da in der nächsten mathearbeit er immer noch
die gleiche lehrerin haben wird.

ZU UNSEREN MEINUNGSVERSCHIEDENHEITEN:
super.

ZU DEN ÄNDERUNGSVORSCHLÄGEN:
ich kann mir nicht vorstellen, daß die tatsache,
daß sie die gleiche lehrerin haben, etwas mit m.´s
anstrengung, begabung oder sonst einem attribut
zusammenhängt.

Pinie composed the next reply as case analyst to the 'constructive critique' of Birke, which lacked new words. Pinie's message contained elicitation with reference to the three social prompts aiming to provoke a response from Birke, who had not yet contributed.

**5th message titled "Antwort auf Kritik" posted by Pinie at 11:10:13**

>> MIT DER ATTRIBUTIONSTHEORIE ERKLÄRBARE
>> FALLINFORMATIONEN:
>> michael und seine eltern machen die mangelnde
>> begabung, die wohl in der familie liegt, für die schlechten
>> matheleistungen verantwortlich - laut b. weiner also eine
>> internal stabile attribution.
>> seine lehrerin 'beschuldigt' ihn der faulheit,
>> eine internal variable attribution.
>> FÜR DIESEN FALL RELEVANTE BEGRIFFE DER
>> ATTRIBUTIONSTHEORIE:
>> - GEHT DER ATTRIBUTION EIN ERFOLG ODER EIN
>> MISSERFOLG VORAUS?
>> misserfolg
>> - IST DIE LOKALITÄT DER ATTRIBUTION INTERNAL ODER
>> EXTERNAL?
>> internal, da die begabung beim schüler selbst
>> lokalisiert ist
>> - WIRD AUF EINE STABILE ODER EINE VARIABLE
>> URSACHE ATTRIBUIERT?
>> auf eine stabile
>> - ATTRIBUIERT DER BETROFFENE SELBST ODER WIRD VON
>> AUßEN FREMD ATTRIBUIERT?
>> beides ist der fall, denn michael selbst sagt,
>> daß mathe nicht "sein fach" ist (hier also die
>> selbstattribution), andererseits liegt auch fremdattribution vor, da
>> seine eltern ihn ebenfalls durch ursachenzuschreibungen
>> beeinflussen .
>> ausserdem attribuiert seine lehrerin von aussen,
>> indem sie sagt, m. wäre faul.
>> PROGNOSE UND PÄDAGOGISCHE KONSEQUENZEN AUS DER
>> PERSPEKTIVE DER ATTRIBUTIONSTHEORIE:
>> michael wird sich wohl kaum weiter anstrengen, da
>> er überzeugt ist, nichts an seinen leistungen im
>> fach mathe ändern zu können. die stabile variable ist
>> ja für ihn nicht veränderbar.
>> NICHT MIT DER ATTRIBUTIONSTHEORIE ERKLÄRBARE

>> FALLINFORMATIONEN:
>> michaels vater hatte ebenfalls untricht bei der
>> gleichen lehrerin, das ist aber meiner meinung nach keine
>> attribution.
> FOLGENDE PUNKTE SIND MIR NOCH UNKLAR:
> PUNKTE BEI DENEN WIR NOCH NICHT EINER MEINUNG
> SIND:
> MEINE VORSCHLÄGE FÜR EINE ÄNDERUNG DER ANALYSE
ZU DEN UNKLARHEITEN:
ist scheinbar alles klar?!

ZU UNSEREN MEINUNGSVERSCHIEDENHEITEN:
sind wohl einer meinung.

ZU DEN ÄNDERUNGSVORSCHLÄGEN:
blebt alles so, wie es ist...?

In the 6<sup>th</sup> message, Birke produced her first and only contribution in this web-based discussion board as response to Pinie's elicitation and the prompts. This response was not at all intended by the prompts.

6th message titled "Konstruktive Kritik" posted by Birke at 11:21:06

>>> MIT DER ATTRIBUTIONSTHEORIE ERKLÄRBARE
>>> FALLINFORMATIONEN:
>>> michael und seine eltern machen die mangelnde
>>> begabung, die wohl in der familie liegt, für die schlechten
>>> matheleistungen verantwortlich - laut b. wäere also eine
>>> internal stabile attribution.
>>> seine lehrerin "beschuldigt" ihn der faulheit
>>> eine internal variable attribution.
>>> FÜR DIESEN FALL RELEVANTE BEGRIFFE DER
>>> ATTRIBUTIONSTHEORIE:
>>> - GEHT DER ATTRIBUTION EIN ERFOLG ODER EIN
>>> MISSERFOLG VORAUS?
>>> misserfolg.
>>> - IST DIE LOKALITÄT DER ATTRIBUTION INTERNAL ODER
>>> EXTERNAL?
>>> internal, da die begabung beim schüler selbst
>>> lokalisiert ist.
>>> - WIRD AUF EINE STABILE ODER EINE VARIABLE
>>> URSACHE ATTRIBUIERT?
>>> auf eine stabile.
>>> - ATTRIBUIERT DER BETROFFENE SELBST ODER WIRD VON
>>> AUSSEN FREMD ATTRIBUIERT?
>>> beides ist der fall, denn michael selbst sagt,
>>> daß mathe nicht "sein fach" ist (hier also die
>>> selbstattribution), andererseits liegt auch fremdattribution vor, da
>>> seine eltern ihn ebenfalls durch ursachenzuschreibungen
>>> beeinflussen.
>>> außerdem attribuiert seine lehrerin von außen,
>>> indem sie sagt, m. wäre faul.
>>> PROGNOSE UND PÄDAGOGISCHE KONSEQUENZEN AUS DER
>>> PERSPEKTIVE DER ATTRIBUTIONSTHEORIE:
>>> michael wird sich wohl kaum weiter anstrengen, da
>>> er überzeugt ist, nichts an seinen leistungen im
>>> fach mathe ändern zu können, die stabile variable ist
>>> ja für ihn nicht veränderbar.
>>> NICHT MIT DER ATTRIBUTIONSTHEORIE ERKLÄRBARE
>>> FALLINFORMATIONEN:
>>> michaels vater hatte ebenfalls untricht bei der
>>> gleichen lehrerin, das ist aber meiner meinung nach keine

```
>>> attribution:
>> FOLGENDE PUNKTE SIND MIR NOCH UNKLAR:
>> PUNKTE BEI DENEN WIR NOCH NICHT EINER MEINUNG
>> SIND:
>> MEINE VORSCHLÄGE FÜR EINE ÄNDERUNG DER ANALYSE:
> ZU DEN UNKLARHEITEN:
> ist scheinbar alles klar?!
> ZU UNSEREN MEINUNGSVERSCHIEDENHEITEN:
> sind wohl einer meinung.
> ZU DEN ÄNDERUNGSVORSCHLÄGEN:
> bleibt alles so, wie es ist...?
FOLGENDE PUNKTE SIND MIR NOCH UNKLAR:
deine analyse ist ok.
PUNKTE BEI DENEN WIR NOCH NICHT EINER MEINUNG
SIND:
ich bin einverstanden
MEINE VORSCHLÄGE FÜR EINE ÄNDERUNG DER ANALYSE:
kann alles so bleiben ,wie es ist
```

The 7<sup>th</sup> message – a constructive critique – was posted by Ahorn at 11:31:07. Ahorn only responded to the last 'modification proposal' prompt. Here, Ahorn aimed to repair the inaccurate use of the term 'attribute' and put the case information excluded by Pinie into the context of Weiner's (1985) attribution theory.

```
7th message titled "Konstruktive Kritik" posted by Ahorn at
11:31:07
>>> MIT DER ATTRIBUTIONSTHEORIE ERKLÄRBARE
>>> FALLINFORMATIONEN:
>>> michael und seine eltern machen die mangelnde
>>> begabung, die wohl in der familie liegt, für die schlechten
>>> matheleistungen verantwortlich - laut b. weiner also eine
>>> internal stabile attribution.
>>> seine lehrerin "beschuldigt" ihn der faulheit,
>>> eine internal variable attribution.
>>> FÜR DIESEN FALL RELEVANTE BEGRIFFE DER
>>> ATTRIBUTIONSTHEORIE:
>>> - GEHT DER ATTRIBUTION EIN ERFOLG ODER EIN
>>> MISSERFOLG VORAUS?
>>> misserfolg
>>> - IST DIE LOKALITÄT DER ATTRIBUTION INTERNAL ODER
>>> EXTERNAL?
>>> internal, da die begabung beim schüler selbst
>>> lokalisiert ist
>>> - WIRD AUF EINE STABILE ODER EINE VARIABLE
>>> URSACHE ATTRIBUIERT?
>>> auf eine stabile
>>> - ATTRIBUIERT DER BETROFFENE SELBST ODER WIRD VON
>>> AUßEN FREMD ATTRIBUIERT?
>>> beides ist der fall, denn michael selbst sagt,
>>> daß mathe nicht 'sein fach' ist (hier also die
>>> selbstattribution), andererseits liegt auch fremdattribution vor, da
>>> seine eltern ihn ebenfalls durch ursachenzuschreibungen
>>> beeinflussen ,
>>> ausserdem attribuiert seine lehrerin von aussen,
>>> indem sie sagt, m. wäre faul.
>>> PROGNOSE UND PÄDAGOGISCHE KONSEQUENZEN AUS DER
>>> PERSPEKTIVE DER ATTRIBUTIONSTHEORIE:
>>> michael wird sich wohl kaum weiter anstrengen, da
>>> er überzeugt ist, nichts an seinen leistungen im
>>> fach mathe ändern zu können, die stabile variable ist
```

> >> ja für ihn nicht veränderbar
> >>> NICHT MIT DER ATTRIBUTIONSTHEORIE ERKLÄRBARE
> >>> FALLINFORMATIONEN:
> >>> michaels vater hatte ebenfalls unrecht bei der
> >>> gleichen lehrerin, dies ist aber meiner meinung nach keine
> >>> attribution.
> >> FOLGENDE PUNKTE SIND MIR NOCH UNKLAR:
> >> Warum ist es keine Attribution des Vaters, wenn
> >> er bei der gleichen Lehrerin Unterricht hatte?
> >> PUNKTE BEI DENEN WIR NOCH NICHT EINER MEINUNG
> >> SIND:
> > Bin völlig deiner Meinung.
> > MEINE VORSCHLÄGE FÜR EINE ÄNDERUNG DER ANALYSE:
> > Letzten Punkt (besser) begründen! WARUM?
> > ZU DEN UNKLARHEITEN:
> > ich dachte,daß es irgendwie nicht mit den
> > attributionen zusammenhängt, es könnte aber auch sein, daß hier
> > das attribut "zufall" vorliegt, was aber nicht
> > wahrscheinlich ist, da in der nächsten mathearbeit er immer noch
> > die gleiche lehrerin haben wird
> > ZU UNSEREN MEINUNGSVERSCHIEDENHEITEN:
> > super.
> > ZU DEN ÄNDERUNGSVORSCHLÄGEN:
> > ich kann mir nicht vorstellen, daß die tatsache,
> > daß sie die gleiche lehrerin haben, etwas mit m. s
> > anstrengung, begabung oder sonst einem attribut
> > zusammenhängt.

FOLGENDE PUNKTE SIND MIR NOCH UNKLAR:


PUNKTE BEI DENEN WIR NOCH NICHT EINER MEINUNG
SIND:


MEINE VORSCHLÄGE FÜR EINE ÄNDERUNG DER ANALYSE:
Ich finde die Zuschreibung (bzw. das Attribut)des
Vaters besteht darin, die Lehrerin mitverantwortlich für
schlechte Zensuren zu machen, da er ja auch hatte,
genauso wie die schlechten Zensuren, oder? ("Der
Schrecken der Mathematiklehrer")
Bin mir aber auch nicht so sicher.

Finally, Pinie composed the final analysis of the case which was basically built on the initial analysis, but also carefully integrated aspects that
Ahorn had brought forward. Pinie no longer responded to the last prompt.

8th message titled "Neue Analyse des Falls Asien" posted by Pinie
at 11:419:34

MIT DER ATTRIBUTIONSTHEORIE ERKLÄRBARE
FALLINFORMATIONEN:
1. mangelnde begabung michaels, die scheinbar in
der familie liegt
2. michaels faulheit
3. der umstand, daß sein vehrhältnis zur lehrerin
quasi durch den vater schon negativ vorbelastet ist

FÜR DIESEN FALL RELEVANTE BEGRIFFE DER
ATTRIBUTIONSTHEORIE:
- GEHT DER ATTRIBUTION EIN ERFOLG ODER EIN
MISSERFOLG VORAUS?
eindeutig michaels misserfolge in mathe
- IST DIE LOKALITÄT DER ATTRIBUTION INTERNAL ODER

EXTERNAL?
zu 1.: internal, da wie schon gesagt die bagabung
in ihm selbst lokalisiert ist
zu 2.: internal, auch die faulheit liegt an/in
ihm selbst
zu 3.: external, es ist eher ein zufall, daß sein
vater die gleiche lehrerin hatte
- WIRD AUF EINE STABILE ODER EINE VARIABLE
URSACHE ATTRIBUIERT?
zu 1.: stabil
zu 2.: variabel, er könnte sich sehr wohl
anstrengen
zu 3.: auch eher variabel, er könnte ja mit der
lehrerin auch zurechtkommen, obwohl sein vater
schwierigkeiten mit ihr hatte
- ATTRIBUIERT DER BETROFFENE SELBST ODER WIRD VON
AUßEN FREMD ATTRIBUIERT?
zu 1.: beides
zu 2.: fremdattribution, da durch die lehrerin
zugeschrieben
zu 3.: beides

PROGNOSE UND PÄDAGOGISCHE KONSEQUENZEN AUS DER
PERSPEKTIVE DER ATTRIBUTIONSTHEORIE:
m. wird sich nicht anstrengen, weil er meint,
seine mangelnde begabung nicht wettmachen zu können.
ausserdem denkt er ja, daß sein misserfolg auch
mit an der lehrerin liegt, und daß kann er sowieso
nicht ändern.

NICHT MIT DER ATTRIBUTIONSTHEORIE ERKLÄRBARE
FALLINFORMATIONEN:

## Social processes

In the ESCOS-discourse, learners were expected to respond to the epistemic prompts and to engage in conflict-oriented consensus building in order to collaboratively construct better analyses and to foster more accurate and multiple perspectives upon the subject matter. The data of the conversational activities show, however, that in one analyst-critic discussion thread, the critic was not at all able to engage in conflict-oriented consensus building. Birke's first message lacked new words and in Birke's second message the prompts were used for *quick consensus building*, which is the exact opposite of the intended response. With respect to the desire for clarity, Birke wrote that Pinie's analysis was "ok;" in response to the difference-of-opinions-prompt Birke signaled agreement; concerning modification proposals Birke responded that "everything can stay as it is."

In the parallel analyst-critic discussion thread, the learners also engaged in *quick consensus building*. Ahorn wrote that she "completely agreed" with Pinie's analysis in response to the prompt "WE HAVE NOT REACHED CONSENSUS CONCERNING THESE ASPECTS." Ahorn, as well as Birke, did not respond to this prompt in the intended way. In response to the prompt "THESE ASPECTS ARE NOT CLEAR TO ME YET" Ahorn aimed to learn from Pinie whether the specific case information that Pinie regarded as "CASE INFORMATION THAT CANNOT BE EXPLAINED WITH ATTRIBUTION THEORY" could nevertheless be explained with attribution theory. This can be regarded as an *elicitation*, as it was suggested by the social script component. In her response, Pinie considered applying attribution theory to this case information. She wrote that "it could be that the attribute 'coincidence' applies here." Pinie also engaged in *conflict-oriented consensus building*, however, when she confirmed her prior estimation about the irrelevancy of this case information. In her second and last critique Ahorn elaborates the ways in which this case information may be explained with attribution theory. Due to the fact that this elaboration opposed Pinie's estimation, Ahorn also engaged in *conflict-oriented consensus building*. Ahorn qualified her statements in a way that can be regarded as *elicitation*: Ahorn wrote, "I believe the attribution [...] is directed towards the teacher, isn't it?". Ahorn further wrote that she was "not so sure" herself. As a result of this discussion thread, Pinie *integrated* this aspect in her final analysis. However, Pinie did not integrate this aspect in accordance to Ahorn's interpretation of the case information, which was adequately based on attribution theory. Instead, Pinie produced some misconceptions when she integrated this aspect into her final analysis (see below).

Overall, the discourse contains a *non-transactive discussion thread* as well as a *more transactive discussion thread*. Birke did not join in to analyze the problem case. Birke's only activity was to accept Pinie's initial analysis. Therefore, Birke's discussion thread can be considered as non-transactive. In Ahorn's discussion thread, some transactive discourse moves were displayed. Ahorn referred to one specific aspect of Pinie's analysis. Ahorn and Pinie began to discuss this aspect, which eventually led to Pinie integrating this aspect into her final analysis.

### Cognitive processes

The ESCOS-discourse begins with the first analysis posted by the case analyst, Pinie. This analysis was supported by the epistemic script component. Pinie immediately *built relations between conceptual and problem space*. Furthermore, Pinie and Ahorn mainly engage in this epistemic activity. The frequency of this epistemic activity in the present case study is in accordance with the quantitative data. The discourse also contains some *non-epistemic activities*. After receiving a message without new words, Pinie aimed to elicit some kind of reaction from Birke. In this discussion thread the learners did not actually advance in the learning task, but merely agreed upon the initial analysis.

Apart from identifying specific epistemic activities, the quality of the learners' analyses can be evaluated. The prompt-supported first analysis posted by Pinie closely matches expert analyses. Pinie appeared to have understood that attribution theory is not about actual characteristics of the learner or the environment, but about what learners and associated people *believe* to be causing success or failure. Pinie composed a concise analysis that includes multiple perspectives. One discussion thread was not helping Pinie to refine her first analysis, because no new words were posted by Birke. Ahorn's critique, however, induced Pinie to adjust her analysis.

When comparing Pinie's initial with her final analysis, indications can be found, however, that this adjustment was a *regression* rather than an improvement of the quality of the initial analysis. In her initial analysis Pinie adequately identifies an internal stable attribution, because "[M]ichael and his parents hold lack of talent [...] responsible for bad performance in math." In her final analysis, Pinie suggested that "lack of talent," "[M]ichael's laziness," and "biased relation to the teacher" are relevant case information with regard to Michael's failure in mathematics. This interpretation of the problem case is inaccurate to some extent. Pinie did not indicate that the protagonist of the problem case ascribes causes for his failure. Pinie was referring to "attributes" instead of "attributions."

How did the quality of Pinie's analyses decrease? Pinie's first message, which was supported by the prompts, contained an adequate analysis.

Apparently, lacking a substantial critique, Ahorn asked for the elaboration of the case information that Pinie regarded to as irrelevant for an analysis based on attribution theory. Ahorn's constructive critique aimed to indicate that "Michael's father attributes the failure to Michael's teacher," which becomes evident in Ahorn's second message (overall 7th message of the discourse). Pinie failed to understand this relation between this case information and attribution theory. In her final analysis she wrote that this "attribution" is "external, because it is a coincidence that the father had the same teacher," and is "rather variable, because [Michael] could also get along with the teacher in spite of his father having had problems with her." These analyses are inadequate with respect to Ahorn's objections and with respect to attribution theory. An adequate interpretation of this case information based on attribution theory and Ahorn's suggestions needed to imply that the father attributes and that the attribution can be classified as external and stable, because Michael cannot choose his teachers himself.

There are indications that these misconceptions arose during Pinie's discourse with Ahorn. During the discussion of this case information, Pinie started to refer to "attributes." It is possible that Pinie failed to comprehend Ahorn's critique in the context of attribution theory. Instead of operating on this case information based on attribution theory, Pinie continued to regard this case information as not explainable with attribution theory, saying that she "cannot see why the fact, that they had the same teacher had something to do with [Michael's] endeavors, talent, or any other attribute." Finally, when Pinie considered and integrated this case information into her final analysis she did so on grounds of misconceptions instead of attribution theory.

*Interpretation of the discourse data*

In the ESCOS-discourse, learners were expected to follow a specific discourse structure according to the social script component equivalent to the SCOS-discourse. The data of the conversational activities show that this was the case. The surface structure, illustrated in figure 7.5.4a, reproduces exactly the prescriptions of the social script component in the ESCOS-discourse. In contrast to the SCOS-discourse, however, this surface structure does not correspond to the actual discourse activities of the learning group (see figure 7.5.4b). The "constructive critics" failed to engage in the specific social modes of co-construction suggested by the social script component. An analysis of the individual messages reveals that the social processes hardly match the analyst-critic discourse structure.

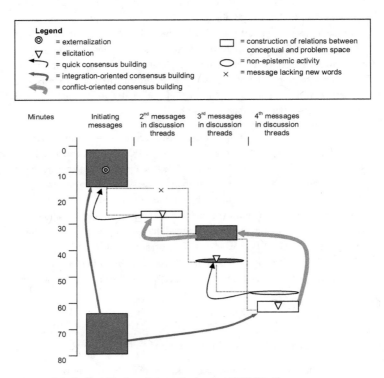

*Figure 7.5.4b*: Graphical coding analysis of an ESCOS-discourse.

In reference to a CSCL approach, which mainly aims to foster coordination between learners, this discussion thread may be superficially regarded to as best practice example for a highly coordinated discourse. The learners had *quickly established consensus*. Apparently, the learners did not suffer from channel reductions of text-based CMC, but could agree quickly on a joint solution. Seen from the socio-cognitive perspectives of educational psychology on collaborative knowledge construction, however, this discussion thread can be referred to as *non-transactive* and can be hypothesized as being unproductive with respect to learning.

Overall, the high quality of the first analysis, possibly induced by the epistemic script component, seemed to impede transactive discourse. The constructive critics failed to refine the initial analysis. The critics could not contribute to the analysis of the case substantially, because they believed that the initial analysis of the case analyst was complete and accurate. Pinie appeared to anticipate critique and was willing to integrate comments by the critics without having built a conceptual space with which the critiques could have been evaluated. As a consequence, Pinie adjusted her analysis in an inadequate way, following misinterpretations of Ahorn's critique and considered the dimensions of attribution theory to also include 'attributes.'

The present ESCOS-case-study can therefore be regarded to as an example for a regressive discourse. What little critique was given to the case analyst is overemphasized and gives rise to modification of an analysis, which was actually fairly adequate in the first place.

# 8 Discussion

In this chapter, the results will be interpreted on the grounds of the conceptual framework of the study and prior findings. Some limitations of the study will be discussed to qualify the results and avoid overgeneralization. The results of the study will be further examined with respect to their significance for the construction of CSCL environments. Finally, future research questions, regarding the implementation of scripts in virtual seminars and the attunement of the instructional support will be discussed.

The rationale of this study was that learners may co-construct applicable knowledge when they collaboratively inquire and discuss problems. Social and cognitive processes of collaborative learners have often been observed, however, as deficient to the ends of learning to apply knowledge. Therefore, both social and cognitive processes may require instructional support. This may be particularly true for CSCL environments, which have been argued to possibly foster collaborative knowledge construction (Clark et al., 2003), but which have also been portrayed as possible barriers for collaborative knowledge construction (Bromme, Hesse, & Spada, 2005). One process-oriented form of instructional support that proved successful in prior studies of educational psychology are cooperation scripts (e.g., Rosenshine et al., 1996). These cooperation scripts have been successfully implemented into CSCL environments with the help of prompts (Fischer, Mandl, Haake, Kollar, 2007; Nussbaum et al., 2002; Scardamalia & Bereiter, 1996). Cooperation scripts may aim at the facilitation of cognitive and social processes. An epistemic cooperation script has been conceptualized to facilitate important cognitive processes and a social cooperation script aimed to facilitate social processes that have been linked to collaborative knowledge construction. The combination of both epistemic and social cooperation scripts was argued to potentially facilitate both social and cognitive processes of collaborative knowledge construction. In other words, the two scripts were meant to foster discourse of learners that is of a high level with respect to epistemic activities and social modes. This means that the scripts aimed to facilitate discourse that can be characterized as task-related, based on theoretical concepts, exploratory, and critical, as learners monitor the adequacy of each other's contributions (cf. Mercer & Wegerif, 1999). Although the discourse of learners may not mirror actual learning processes without loss and distortion of information, a multi-dimensional, qualitative and quantitative discourse analysis may still indicate, to some extent, how knowledge is collaboratively constructed. It was argued that even though the influence of learners' discourse on knowledge acquisition may be highly complex and indirect, social and cognitive processes as they surface in discourse may be related to learning (Gerstenmaier & Mandl, 1999; Vygotsky, 1986).

The results of the study show that cooperation scripts can fundamentally change discourse of learners and can be carefully directed towards specific processes of collaborative knowledge construction. Furthermore, cooperation scripts proved to substantially affect learning outcomes. Results show that all experimental groups acquired knowledge, as compared to their pre-test results, but knowledge acquisition can be further enhanced or impeded by specific CSCL scripts. Therefore, as can be seen in prior studies, this study shows that with the help of prompts, the script approach can be realized in CSCL environments with comparable efficiency (Rosenshine et al., 1996). Furthermore, it could be shown that social and cognitive processes, as they surface in the discourse of learners, do, to some extent, relate to learning outcomes. These results need to be discussed and interpreted in more detail.

## 8.1    Instructional Support by Social Cooperation Scripts

First of all, effects of social cooperation scripts will be discussed with respect to the individual research questions of the study. These research questions were oriented towards effects on social processes, effects on cognitive processes, and effects on outcomes of collaborative knowledge construction.

### 8.1.1 Effects of social cooperation scripts on social processes

To start off, the treatment check shows that the social cooperation script has any one member of a learning group producing exactly the *same amount of messages*. Participants without social cooperation script sent more messages, but show a substantially higher heterogeneity in respect to number of messages. That means, that some learners may produce many messages in discourse without social cooperation script, but some other learners of the same learning group may only post very few messages. Furthermore, the social cooperation script clearly influences the social modes of co-construction. It produces substantial effects on externalization, quick consensus building, and conflict-oriented consensus building. The effect on externalization was large and negative, whereas the social cooperation script produces medium-sized positive effects on quick and conflict-oriented consensus building. The social script affects learners to reduce *externalization*, meaning not to think aloud in front of the group on their own, but rather to orient their contributions towards the contributions of the learning partners. Thus, it can be said that social cooperation scripts can improve transactivity of learners' discourse. The positive effect on *quick consensus building* may indicate that the demands of the social script to critically evaluate contributions of learning partners in a specified time frame are often responded to in an unintended way. It is possible that learners confronted with the demand to be critical, are inclined to signal their general acceptance in spite of their roles in order to foster social coherence. At the same time however, the social script facilitates *conflict-oriented consensus building*, meaning that learners are supported to evaluate contributions from their learning partners more critically and to anticipate critical review regarding their own contributions, which may improve collaborative knowledge construction. Thus it can be stated that the social script works in the intended way to foster a more critical discourse of learners. A possible explanation for facilitating both quick and conflict-oriented consensus building can be deduced from the SCOS-case-study. Learners may engage in 'yes-but-critique,' stating, for instance, that they understood the point of view of their learning partners, but nevertheless formulate critique. Even though social script learners may still feel inclined to soften their critique by signaling quick consensus, the social cooperation script suggests conflict-oriented consensus building to be a socially desirable way to interact with one's learning partners. Conflict-oriented consensus building may often be neglected in CMC, due to high coordination demands. Therefore, facilitation of conflict-oriented consensus building may be particularly important in CSCL environments. Some studies suggest that learners may not often show this social mode in collaborative knowledge construction spontaneously or may not be able to productively resolve conflicts (e.g., Tudge, 1989). In this way, social cooperation scripts can facilitate collaborative knowledge construction and reduce the known satisficing problem of CSCL, meaning that learners tend to *satisfy only minimal requirements* of the collaborative learning environment (Chinn et al., 2000; Linn & Burbules, 1993).

### 8.1.2 Effects of social cooperation scripts on cognitive processes

In general, the cooperation scripts facilitate epistemic and reduce non-epistemic activities. The relative frequency of the individual epistemic activities may differ substantially depending on the actual domain and task of learners (Fischer et al., 2002). It can clearly be seen that the facilitation of epistemic activity supports findings of prior studies, indicating that cooperation scripts realized in CSCL environments reduce non-epistemic and facilitate epistemic activities (Baker & Lund, 1997). Furthermore, the social script in particular reduces the *construction of conceptual space*. This indicates that the social structure provided by the social script demands learners to apply theoretical concepts to problem space rather than to report about them. It is possible that the roles of case analyst and constructive critic are understood in the sense that these roles demand working on the problem, which has been interpreted by the learners as to apply concepts to case information. Different roles, like teacher and student, may instead facilitate the construction of conceptual space (Reiserer, 2003). This assumption is affirmed, because the social script additionally facilitates the *construction of relations between conceptual and problem space*. The social script roles appear to suggest that learners deal with the problem case on grounds of theoretical concepts.

### 8.1.3    Effects of social cooperation scripts on outcomes of collaborative knowledge construction

Collaborative knowledge construction aims, in conclusion, to improve learning outcomes, meaning in particular, to foster applicable knowledge and multiple perspectives on subject matters. Some studies indicate that these goals can rarely be accomplished, because learners rarely show activities spontaneously in collaborative knowledge construction related to these learning outcomes. Thus, the benefits of collaborative knowledge construction are rarely exploited. The social script has aimed to foster social processes in order to actually realize benefits of collaborative knowledge construction. The results show that the social script affects neither focused nor multi-perspective applicable knowledge as co-construct. There is also no social script effect on individually acquired focused applicable knowledge. However, the social script substantially facilitates the *individual acquisition of multi-perspective applicable knowledge* with medium-sized effects. This result could be regarded to as encouraging, because the social script seems to promote the benefits of collaborative knowledge construction, namely to learn to approach a problem from multiple perspectives. The social cooperation script may emphasize the notion of collaborative knowledge construction as a trial and error test bed for learners (Gordin, Gomez, Pea, & Fishman, 1996; Slavin, 1992). Learners may try and err regarding knowledge as co-construct, but this trial and error within a learning group may facilitate individual knowledge construction – particularly when a social cooperation script personalizes mutual monitoring of co-construction processes of peers by realizing such social roles as constructive critic (cf. Schwarz, Neuman, & Biezuner, 2000). Thus, knowledge as co-construct of learning groups supported with a social script may be similar to knowledge as co-construct of control groups, but simultaneously individual knowledge acquisition can be fostered (Larson et al., 1985). Social script learners expect to be criticized by their peers. Thus, learners anticipate socio-cognitive conflict and develop multiple perspectives to explain case information. In summary it can be said that the social script establishes a specific transactive, critical, and error tolerant micro culture of collaborative knowledge construction. The social script facilitates the individual acquisition of multiperspective applicable knowledge and enables learners to apply concepts to a problem from multiple perspectives.

## 8.2 Instructional Support by Epistemic Cooperation Scripts

In this section, effects of epistemic cooperation scripts will be discussed with respect to the individual research questions of the study regarding effects on social processes, effects on cognitive processes, and effects on outcomes of collaborative knowledge construction.

### 8.2.1 Effects of epistemic cooperation scripts on social processes

With regard to social modes of co-construction, the study shows that opposite to the social script, the epistemic cooperation script strongly suggests that learners engage in the social mode of *externalization*. The epistemic script affects learners to think aloud in front of the group, e.g., putting forth suggestions to analyze the cases. Therefore, epistemic scripts may be functional to some extent. It could be argued that groups working on a problem together may be efficiently supported with epistemic cooperation scripts, because they could motivate learners to externalize their knowledge in CMC. Epistemic cooperation scripts, however, appear to have major negative effects on social modes of co-construction for groups who intend to construct knowledge collaboratively. A substantial, medium-sized negative effect of the epistemic script has been found regarding *elicitation*. Apparently, the epistemic script induces learners to analyze the problem cases individually and does not suggest using the learning partners as resources. In this respect, epistemic cooperation scripts could obstruct the benefits of collaborative knowledge construction by reducing the need to inquire and discuss a problem in transactive discourse. Epistemic cooperation scripts may instead distract learners from referring to each other – a problem discussed early on in the scripted cooperation approach (Dansereau, 1988).

This means that the epistemic script may suggest individual approaches to solve a problem. Similar to a checklist, the epistemic script provides learners with an adequate strategy with which they can solve problem cases individually. This 'checklist' may have reduced the motivation of learners to invest a lot of cognitive and meta-cognitive efforts in the actual collaboration. Why should learners, who dispose of a functional and acknowledged strategy, invest extra effort in conflict-oriented, exploratory talk? Epistemic script learners are held to believe they can solve problems alone. Learner may fall victim to an *illusion of competence*, disregarding inconsistencies within their individual approaches. As a consequence, the epistemic script promotes the often-described tendency of learners to minimize collaborative demands (cf. Clark & Wilkes-Gibbs, 1986; Renkl & Mandl, 1995). By providing learners with 'the adequate task strategy,' the epistemic script seems to enable learners to segment and share the task and solve it fairly independent of support by peer learning partners. The technical aspects of CMC may further intensify this effect. Text-based CMC may increase the demands to coordinate communication within a group due to filtered out context cues (cf. Hesse et al., 1997). CMC may therefore foster the tendency to segment and share the task instead of collaborative approaches to critically discuss multiple perspectives. Simultaneously, social relations and responsibility of the anonymous peer learners in CMC may become salient only after some time of collaboration within the group (Matheson & Zanna, 1990). Therefore, epistemic cooperation scripts in CSCL environments may further hamper transactivity in learners' discourse. Less transactive learning discourse may, however, reduce the potential of collaborative learning scenarios to teach learners to adopt and apply multiple perspectives (Teasley, 1997). This may be the case for the epistemic script of this study.

### 8.2.2 Effects of epistemic cooperation scripts on cognitive processes

The epistemic script facilitates learners to *construct problem space* and to *construct relations between conceptual and problem space*. These medium-sized effects of epistemic cooperation scripts on epistemic activities appear to be highly dependent on the single aspects that the epistemic cooperation script aims at. For instance, the epistemic script of this study prompts learners to construct problem space (e.g., with the prompt "CASE INFORMATION THAT CAN/CANNOT BE EXPLAINED WITH AT-TRIBUTION THEORY:") and to produce relations between conceptual and problem space (e.g., with prompts like "DOES A SUCCESS OR FAILURE PRECEDE THIS ATTRIBUTION?"). Thus, the results suggest that cooperation scripts may be carefully directed to influence specific epistemic activities. In this way, the effects of the epistemic script on epistemic activities can be closely related to what the individual prompts of the epistemic script aim at. This systematic facilitation of specific epistemic activities may go beyond a general beneficial effect of cooperation scripts on epistemic activities. The epistemic script of the study is clearly restricted to facilitate construction of problem space and construction of relations between conceptual and problem space in contrast to other epistemic scripts that also facilitate the construction of conceptual space (e.g., Dufresne et al., 1992). As the results show, a general effect on epistemic activities may not only be achieved by a cooperation script directly aiming at specific epistemic activities, but also by scripts that provide structure to social aspects of collaborative knowledge construction or that simply aim to coordinate and reduce demands of computer-mediated discourse (Baker & Lund, 1997).

### 8.2.3    Effects of epistemic cooperation scripts on outcomes of collaborative knowledge construction

The epistemic script fosters *focused applicable knowledge as a coconstruct*. This major effect indicates that epistemic cooperation scripts may direct the attention of learners in specific directions and facilitate learners to apply knowledge to the more central aspects of a problem. There are indications that the epistemic script enables learners to concentrate on the most elementary aspects of the complex problem cases and solve them as long as they are provided with this kind of support in the collaborative phase. Multi-perspective applicable knowledge as co-construct, however, is not influenced by the epistemic script. This result emphasizes that the epistemic cooperation script of this study strongly guides cognitive processes of learners to concentrate on specific aspects of the task. In this way, the epistemic script may not sufficiently facilitate a complete analysis of the problem cases, which should include multiple perspectives. Instead, the epistemic script suggests that an analysis is completed once any prompt has been responded to one time. In this regard, the epistemic script does not facilitate recurrent analyses from multiple perspectives in discourse.

The facilitation of focused applicable knowledge as co-construct indicates that learners are enabled to identify the most elementary aspects of the complex problem cases, but also perceive the demands of the learning task to be satisfied when these aspects are identified with help of the epistemic cooperation script. There is no meta-cognitive component of the epistemic script to facilitate intra-personal monitoring of learners to determine whether a first analysis was adequate, sufficient, and actually complete (cf. King, 1999). Therefore, the epistemic script may not have instantiated a more expertlike approach towards solving the problems, which is characterized by first constructing conceptual, instead of problem space (Dufresne et al., 1992). The epistemic script may in this way pose an efficient tool for experts who already possess the relevant knowledge structures, but guide novice learners to focus on problem space. As was shown above, the epistemic script also reduces transactivity, which may impede learning partners to inter-personally monitor the adequacy of the individual analyses as well.

Furthermore, the results regarding the learning outcomes show that the epistemic script impedes the individual acquisition of applicable knowledge. Even though there is no substantial effect of the epistemic script on the individual acquisition of focused applicable knowledge, there is a strong and significant negative effect of the epistemic script on the *individual acquisition of multi-perspective applicable knowledge*. This effect is opposite to the effect of the social cooperation script regarding this learning outcome. Even though learners may be able to solve the problem cases well, as long as they are provided with the epistemic cooperation script, they fail to acquire knowledge that can be applied individually and independent of the epistemic cooperation script. It is possible that the epistemic cooperation script reduces or substitutes reflective reasoning in regard to the learning content (Reiser, 2002). Similar findings have been reported by Webb et al. (1986), who argue that collaborative tasks accomplished with instructor prompting do not facilitate knowledge construction, because students may depend on the instructor's assistance rather than to learn from the instructor's suggestions and prompts. The ECOS may foster the identification and solution of problems, but does not facilitate learners to construct a mental conceptual model to comprehend various facets of the problem and the application of the relevant theoretical concepts.

If these explanations apply, epistemic cooperation scripts could be improved in several ways. First, epistemic scripts may *better represent expert strategies for learners* (Dufresne et al., 1992). The rationale of the epistemic script was to support learners to find the relevant case information and to distinguish it from irrelevant case information. Apparently, this does not represent expert strategies well. Epistemic script should maybe instead guide learners' attention towards conceptual space. Second, *meta-cognitive components* may need to be added to epistemic scripts, e.g., to prompt learners to consider the improvement of their individual knowledge (King, 1999; Larson et al., 1985), or *social components* as has been accomplished in the group with both epistemic and social script (see also Palincsar & Herrenkohl, 1999). Third, successively *fading* this kind of instructional support (Collins et al., 1989) may facilitate the internalization of task strategies to apply knowledge to problem case information suggested by epistemic cooperation scripts.

Thus, it can be summarized that where the social cooperation script exploits the benefits of collaborative knowledge construction, the epistemic cooperation script, in contrast, impedes the specific potential of collaborative knowledge construction. Learners are facilitated to apply knowledge as co-construct, but are affected negatively to acquire multiple perspectives of how to apply knowledge individually.

## 8.3 Interaction Effects of the Factors 'Social Cooperation Script' and 'Epistemic Cooperation Script'

The rationale of the study was, on one hand, to facilitate learners to deal with the tasks (epistemic activities), and, on the other hand, to facilitate learners' interactions (social modes of co-construction). The combination of both cooperation scripts was supposed to result in a discourse of high quality regarding epistemic activities and social modes of co-construction. Basically, additive effects have been expected on grounds of the socio-cognitive approaches. However, the factors, 'social script' and 'epistemic script,' interact in some respects. The combination of both scripts sometimes produces results beyond those that could be achieved with the scripts separately or the scripts seem to counteract and nullify each other. In this way, several explanations may apply to the single interaction effects and need to be considered against the theoretical background and the respective context of each interaction effect found.

### 8.3.1 Interaction effects on social processes

With respect to social modes of co-construction, a substantial and large interaction effect could be found regarding *externalization*. Where the epistemic script produces a large positive effect on externalization, the social script substantially reduces externalization in discourse of collaborative learners. Thus, an actual nullification could be expected by the combination of both cooperation scripts. However, another explanation of interaction effects seems to apply, namely that one of the factors appears to be *dominant* regarding a specific dependent variable. In the condition with both scripts, the case analyses are facilitated by the epistemic script component, but simultaneously, distribution of roles and structuring of specific activities of these roles are facilitated throughout discourse by the social script component. In this way, the epistemic script component is actually reserved to the case analyst of the social script. Both constructive critics are to relate to the initial case analysis and not to apply the epistemic script component themselves. Therefore, the factor 'social script' dominates the factor 'epistemic script' regarding the social processes of collaborative knowledge construction. The combination – high quality of initial analysis and transactive discussion of this analysis – aims to result in interaction effects regarding processes and outcomes of collaborative knowledge construction that could not be achieved with social and epistemic cooperation scripts separately. The case studies suggest, however, that learners may not be able to act out the social role of constructive critic when an initial analysis is of a high quality. In this case, the basic premises of collaborative knowledge construction may lack. Collaborative knowledge construction requires complex tasks and resource interdependence (Cohen, 1994). When facilitated with the epistemic prompts, the collaborative task to discuss a complex problem may have shifted to become a more simple task that can be solved by an individual learner.

### 8.3.2     Interaction effects on cognitive processes

With respect to epistemic activities, a medium-sized interaction effect on the *construction of relations between conceptual and problem space* can be found. As both social and epistemic cooperation scripts facilitate the construction of relations between conceptual and problem space, it might be expected that these effects add up. The results suggest a plateau effect, however. That means, that each of the script conditions facilitates construction of relations between conceptual and problem space to a certain upper limit.

Even though proportional differences are slim, it has been shown that groups facilitated with any kind of script or combination of scripts perform better than the control group with respect to the total amount of constructions of relations between conceptual and problem space. This indicates that the amount of relations between conceptual and problem space levels off. It is possible that an optimal level can be achieved with any form of the researched cooperation scripts (see also Baker & Lund, 1997).

### 8.3.3 Interaction effects on outcomes of collaborative knowledge construction

No interaction effects can be found with respect to focused and multi-perspective applicable knowledge as co-construct or with respect to individual acquisition of multi-perspective applicable knowledge. A large interaction effect on the acquisition of focused applicable knowledge is observable, however. Neither treatment appears to separately foster acquisition of focused applicable knowledge. When both cooperation scripts are combined, however, the level of individual acquisition of focused applicable knowledge can be improved. Then again, this level does not supersede the level of the control group. Learners who are provided with both scripts reach a similar level, regarding this outcome variable as the control group.

This substantial interaction effect towards the opposing direction as the main (non-substantial) effects of the two factors, which makes the groups with both scripts more similar to the control groups, could be explained by an *overprompting* or *overscripting* effect (Dillenbourg, 2002; Rosenshine et al., 1996). The two scripts may direct the attention of learners towards too many points and thus strain the cognitive load of learners (cf. Dansereau, 1988; Gräsel, Fischer, & Mandl, 2001; Gräsel et al., 1994). Therefore, learners of the group with both scripts may abandon the scripts altogether, that is they may ignore the scripts. There are no further indications, however, that learners cannot use both cooperation scripts simultaneously. The treatment check shows that learners of all experimental conditions can sufficiently handle the CSCL environment in any of the script conditions. Thus, another explanation needs to be applied. It is possible that the epistemic script component facilitates focused knowledge as co-construct, and simultaneously the social script component facilitates transfer and internalization of focused knowledge as co-construct. Thus, learners supported with both scripts may individually acquire focused applicable knowledge. Here, further analyses and studies may be necessary to evaluate and eventually differentiate whether learners were facilitated by both scripts to apply and then to transfer co-constructed knowledge beyond the collaborative phase.

## 8.4    Relations between Processes and Outcomes of Collaborative Knowledge Construction

Social and epistemic cooperation scripts indeed influence the *discourse processes* of learners and – *eventually* as a consequence – also affect the *outcomes of collaborative knowledge construction*. Against the theoretical background of cooperation scripts, this assumed relation may be independent of the respective cooperation scripts. Groups without cooperation scripts that are induced externally by interface design, act based on the internally represented scripts they themselves spontaneously apply. For instance, learners of the control group may eventually decide to distribute the responsibility of the three cases equally within the groups of three. However, this was not evident in this study. Maybe the database needs to be enlarged in order to quantitatively find discourse patterns of unscripted groups (Neuman, Leibowitz, & Schwarz, 2000).

Of course, no relation between processes and outcomes may be causal. Learners with superior individual learning prerequisites, for instance, may be able to show specific processes of collaborative knowledge construction and simultaneously acquire knowledge due to their superior prior knowledge base rather than because they may have discussed subject matters in a specific way (cf. Renkl, 1997). Furthermore, social and cognitive processes may not be assessed without distortion in discourse. Learners may not actively engage in discourse activities but still have advantageous learning prerequisites and construct knowledge successfully (Fischer & Mandl, 2000).

In the following paragraphs, results will be interpreted against the background of socio-cognitive approaches holding the assumption that social and cognitive processes affect outcomes. It should be carefully noted, however, that the quantitative analysis of these relations is exploratory and that further qualitative case studies should be regarded to replicate these results. In order to examine how the social and the cognitive processes relate to individual acquisition of applicable knowledge, the following analyses have been made. First, the relation between social processes and individual acquisition of applicable knowledge, and second, the relation between cognitive processes and individual acquisition of applicable knowledge has been analyzed and will be discussed in this section. Finally, relation between knowledge as co-construct and individually acquired knowledge will be explored. These relations have been analyzed on an individual level in order to analyze whether or not specific activities can be related to learning outcomes of individual learners.

### 8.4.1      Relations between social processes and outcomes of collaborative knowledge construction

There is a range of approaches linking various social modes of co-construction to individual acquisition of knowledge. The idea of transactivity offers a framework for how these social modes may be weighted with regard to collaborative knowledge construction. However, any social mode has specific functions in the discourse of learners. Therefore, quantitative relations between social modes of co-construction and individual knowledge acquisition need to be carefully interpreted.

With respect to the relation between social modes of co-construction and individually acquired focused applicable knowledge, *conflict-oriented consensus building*, in particular, appears to be important. Against a socio-cognitive background, learners who critically evaluate the contributions of their learning partners may also need to develop their own mental models and acquire new knowledge to sustain critique (Walton & Krabbe, 1995). Conflict-oriented consensus building may require learners to explore the learning environment more actively than other social modes of co-construction. Thus, with respect to the relation between social modes of co-construction and individual acquisition of focused applicable knowledge, the results are consistent with socio-constructivist assumptions about conflict-oriented consensus building as a social mode that is important to learning (De Lisi & Goldbeck, 1999; Piaget, 1985).

*Individual acquisition of multi-perspective applicable knowledge* relates to *elicitation*. This result emphasizes King's (1999) findings on the importance of asking questions for collaborative knowledge construction. It is possible that learners who are using learning partners as a resource may be more able to add their partners' perspectives to their own. Thus, elicitation may indicate that learners are accumulating multiple perspectives.

Both relations of these social modes with the respective dimensions of individually acquired applicable knowledge are substantial, but are of small effect size. These relations may indicate that individual social modes may be hardly linkable to the learning outcome of individual learners. Considering that any social mode may have a specific function for collaborative knowledge construction, analysis may need to shift to questioning, to what extent any particular social mode is actually functional in the specific context in learners' discourse. For instance, does conflict-oriented consensus building actually result in accommodation of the individual perspectives, or do learners instead defend their positions eristically? Discourse structures may need to be analyzed in order to find indications that any particular social mode leads to the expectations of what is to be theoretically assumed.

**8.4.2     Relations between cognitive processes and out-
comes of collaborative knowledge construction**

It has been argued that epistemic activities, in contrast to non-
epistemic activities, relate in general with individual acquisition of applicable knowledge in collaborative knowledge construction, which is based on
complex problem-solving tasks (Cohen, 1994). Recent studies refined this
perspective, indicating that epistemic activities characterized by the use of
theoretical concepts and their relations to case information relate to individual knowledge acquisition (Fischer et al., 2002), whereas the equally epistemic activity of working on case information does not relate to knowledge
acquisition. The results of the present study can affirm this relation. Individually acquired focused applicable knowledge clearly relates to a model
consisting of both the construction of conceptual space and the construction
of relations between conceptual and problem space, even though the construction of relations between conceptual and problem space explains most
of the variance. Regarding relations of epistemic activities with individual
acquisition of multi-perspective applicable knowledge, construction of relations between conceptual and problem space is the only epistemic activity
showing co-variation. These findings indicate that applicable knowledge in
particular may be facilitated by application of concepts rather than reporting
on concepts. The construction of relations between conceptual and problem
space takes on a pivotal role in several respects. Learners have mainly engaged in this epistemic activity. Furthermore, large effect sizes regarding
the relations between this epistemic activity and outcomes are observable. It
is possible that this epistemic activity may indicate, in a refined manner,
whether or not the learners engaged in the *actual* learning task to analyze
problems. Therefore, Cohen's (1994) perspective can be confirmed and refined. Learners dealing with complex learning tasks basically *do* acquire
knowledge when they remain on task.

### 8.4.3 Relations between knowledge as co-construct and individual acquisition of knowledge

The two outcome dimensions, knowledge as co-construct and individually acquired knowledge, hardly relate. Only multi-perspective knowledge as co-construct co-varies to some extent with individual acquisition of focused applicable knowledge. This result is counter-intuitive to some extent. Against a socio-cultural background, relations between knowledge as co-construct and individual acquisition of knowledge, should be strong. Learners who collaboratively apply focused knowledge should learn how to apply this knowledge individually as well. There is, however, no indication that this is the case. Presumably, the knowledge as co-construct is not subject to internalization and transfer, meaning that the collaborative knowledge construction may rather be referred to as a test bed in which participants may learn from their mistakes rather than by adequate knowledge application. The relation between multi-perspective knowledge as co-construct to individually acquired focused knowledge may affirm this assumption. Learners who apply knowledge from multiple perspectives in collaboration may better conceptualize the basic ideas of the theory they are supposed to apply. In this respect learners individually acquire focused knowledge.

## 8.5 Discussion of Processes and Discourse Structures in the Case Studies

The case studies have aimed to illustrate the quantitative results. Furthermore, the case studies aimed to identify the social and cognitive discourse structures as well as the comprehension failures of students. The case studies have added value to the study by identifying and illustrating these aspects in the separate experimental conditions. Quantitative analyses may conceal how and why social and cognitive processes emerge in discourse of learners. Case studies reveal to what extent specific process phenomena correspond with theoretical assumptions. First, discourse structures will be discussed in reference to each of the case studies. Subsequently, typical comprehension failures will be examined as they have emerged regardless of experimental condition.

### Discourse structures in the case studies

The first case study describes a learning group that was not facilitated with any script. The DWS-discourse does not span over the whole collaborative phase. Instead, the learners have quickly built consensus satisficing minimal requirements of the learning environment. Satisficing has been regarded to as a pivotal problem of collaborative knowledge construction (Chinn et al., 2000). The reasons for the satisficing discourse may be that students generally aim to minimize their efforts. It is possible that students have adopted a learning culture in which resources need to be economized. For instance, students may frequently experience that they need to concentrate only on tests that determine their further certification. Therefore, learners may not spend more effort than is necessary on learning environments that are not directly related to certification. There are indications, however, that oppose this explanation. In general, students have indicated in questionnaires and debriefings that they were interested in the learning environment. The participants of the DWS-case-study pose no exception to this generally high interest. Therefore, an alternative explanation needs to be applied. The discourse about the problem case took place in the last few minutes of the collaborative phase. In spite of the fact that the learners had an online timer as well as a physical clock in order to manage time, it is possible that the participants were not able to coordinate the three problem cases well. It is also observable that the learners in the DWS-case-study needed to coordinate themselves throughout the discourse. Therefore, learners may have had difficulties in scripting their collaborative activities themselves. This problem may have been increased by the fact that collaboration needed to be coordinated via text-based CMC. Therefore, it could be said, that access to CMC may not be enough to foster collaborative knowledge construction. Instead, the DWS-case-study may pose a good example for the need to consider consolidated findings of CSCL research. CMC interfaces may not be purposefully designed for CSCL and can be substantially improved by inducing structure to coordinate learners' interactions (cf. Hesse et al., 1997).

The second, SCOS-case-study showed, that collaborative knowledge construction could be facilitated in CSCL to produce specific discourse structures. The SCOS-discourse primarily reflects a highly transactive discourse structure. Transactivity has been regarded as a general measure of discourse quality with respect to collaborative knowledge construction (Teasley, 1997). The SCOS-case-study, in comparison to the quantitative results, suggests that transactivity may be a concept, which needs to be further refined. The social modes of co-construction may have separate functions for collaborative knowledge construction. Furthermore, discourses may need to be additionally analyzed regarding epistemic activities. The learners of the SCOS-case-study have continuously engaged in transactive discourse, but only some specific transactive interactions have added value to collaborative knowledge construction. In one of the discussion threads, the case analyst was able to adequately construct conceptual space together with one of the constructive critics. In another discussion thread, however, the second constructive critic did not produce valuable input for the case analyst to refine her analysis of the problem case. Therefore, a decisive aspect of the SCOS-case-study was the ability of the case analyst to evaluate the qualities of the critiques and to flexibly apply the social cooperation script.

Prototypical for problematic effects of the epistemic cooperation script was the ECOS-case-study. The three participants posted three individual analyses of the problem case without referencing each other. Even though some learners' analyses of the problem cases were adequate in comparison to an expert solution, some other learners produced analyses based on misconceptions of the attribution theory. It could be argued that these learners were not able to exploit the specific advantages of collaborative knowledge construction, because they did not interact with each other actively. In this way, epistemic cooperation scripts may withdraw the basis for collaborative knowledge construction. In this way, learners may also not participate equally and share conceptions about a subject matter. This could be regarded to as particularly problematic, because one of the implicit goals of collaborative knowledge construction is to establish equality in class-rooms and foster shared knowledge in learning groups (Cohen & Lotan, 1995; Fischer & Mandl, 2001).

Finally, the ESCOS-case-study offers some explanation for why the combination of a social script that improved interactions and an epistemic script that improved problem-solving rarely added up to facilitate collaborative knowledge construction. The social script component warranted that learners engaged in analyst-critic discourse structures. Therefore, the ES-COS-discourse was more transactive in comparison to the discourse without the social script component. In addition, the epistemic script component in the ESCOS-case-study supported the case analyst to compose a highly adequate analysis. In this way, however, the constructive critics who were supposed to help to refine the initial analysis were confronted with a specific problem; they could not adopt their social roles well, because the initial analysis left little space for improvement. As a consequence, the case study poses a good example of how collaborative learners may regress to some extent, because in collaborative knowledge construction, adequate conceptions may also be called into question (see Schwarz et al., 2000).

In conclusion, learning environments that demand learners to discuss complex subject matters in a zone of proximal development can be referred to as a fundament for collaborative knowledge construction (Vygotsky, 1978). The learners are supposed to explore subjects that they are not yet familiar with. Pivotal to collaborative knowledge construction in this zone of proximal development is the instructional support. Those case studies, in which an epistemic script component was applied (ECOS and ESCOS), suggest that this instructional support may facilitate learners' problem-solving. The case studies further suggest, however, that an epistemic script component may reduce the quality of interactions. Learners who are equipped with the one correct task strategy to apply knowledge to problem case information do not need to engage in collaborative knowledge construction.

In comparison to the quantitative results, the case studies provide information about how and why specific discourse phenomena may be more or less beneficial for collaborative knowledge construction. The quantitative results only explain the frequency of specific process phenomena. The case studies, however, identify specific process phenomena and their interplay in specific discourse structures. Furthermore, the case studies indicate that scripts may achieve facilitation of specific processes at some costs. For instance, social scripts may not improve the quality of learners' analyses and at the same time may foster an illusion of competence due to the authority of the externally induced social roles. Epistemic scripts may facilitate learners to build adequate analyses at the costs of social interactions and internalization of applicable knowledge. The combination of both scripts may not add up the benefits of both social and epistemic script. An initial analysis, which is supported by the epistemic script component, may impede the functionality of the social script component. It must be possible to make improvements upon the initial analysis in order to be able to act out the social role of 'constructive critic'. The epistemic script component, however, may facilitate 'case analysts' to compose highly adequate initial analyses. Therefore, it can be generalized from the case studies that scripts can successfully aim at specific processes, but may also produce uninvited 'side effects.'

### Comprehension failures

The case studies revealed typical comprehension failures of attribution theory. It is important to discuss these comprehension failures as indicators for the general understanding of learners in the individual case studies. Furthermore, the identification of comprehension failures may be important in order to improve learning environments in the social sciences. Teachers may anticipate the specific comprehension failures of students and facilitate learning with the aim to reduce these misconceptions. Weiner's (1985) attribution theory may pose a typical cognitive theory that focuses on cognition and beliefs of individuals as powerful factors for learning, rather than actual personality traits. Learners often fail to understand that attribution theory does not reveal actual characteristics of a situation but instead explains assumptions of individuals. The case studies show that a cognitive perspective may not be intuitively comprehendible for students. Students may not be accustomed to operating with cognitive concepts. Therefore, students of the social sciences may need to be additionally facilitated to comprehend the underlying assumptions of cognitive theories.

## 8.6      Summary of the Results

The results of the study offer a rather coherent picture of the processes and outcomes, as well as the collaborative knowledge construction in CSCL environments and their facilitation. Cooperation scripts may facilitate the actual benefits of collaborative knowledge construction, which are to learn to apply knowledge from multiple perspectives. Cooperation scripts may also bring forward *overscripting effects*, impeding transactivity and obstructing collaborative knowledge construction. Some results correspond to prior findings on the barriers of collaborative knowledge construction. Students who aim to only meet minimum requirements with regard to the learning task and who refrain from transactive forms of collaboration, may work together sufficiently, but acquire less applicable knowledge individually (Hogan et al., 2000; Linn & Burbules, 1993). CSCL scripts may facilitate specific discourse activities and learning outcomes.

*Social cooperation scripts* can motivate learners to engage in a more conflict-oriented social mode in spite of the possibility that this mode may be socially undesirable. Even though, social scripts may make the learning task more difficult and demanding to some extent, this extra effort might be decisive for the individual acquisition of knowledge (Reiser, 2002; Wiley & Voss, 1999).

Some forms of instructional support may overprompt or overscript learners, thus easing the learning tasks in an exaggerated manner, meaning that the complexity of the task is being substantially reduced. Therefore, collaborative knowledge construction that results from a demanding discussion of complex problems may be hampered (Dillenbourg, 2002; Rosenshine et al., 1996). This appears to have been the case regarding the epistemic cooperation script.

*Epistemic cooperation scripts* may impede some important social processes of collaborative knowledge construction. Some learners may focus on content-aspects of the learning task rather than to engage in transactive discourse (Dansereau, 1988; Teasley, 1997). In this way, epistemic scripts may suggest more individual approaches to the learning task. Epistemic scripts may make the learning task too easy increasing overconfidence of learners. Simultaneously epistemic scripts may substitute important meta-cognitive learning activities. Neither the learners themselves nor their learning partners may monitor the adequacy of the individual analyses. Thus, learners may achieve adequate analyses as long as they are supported with epistemic scripts, but may have difficulties internalizing a mental model underlying the epistemic script. In this way, epistemic scripts without any meta-cognitive components may act as a *crutch* for operating with problems, which learners become dependent on, but not as a *scaffold* for knowledge acquisition which is faded out after the learning phase.

## 8.7    Implications of the Study for the Facilitation of CSCL and for further Research Questions

The study yielded that it is possible to facilitate the cognitive and the social processes and the outcomes of collaborative knowledge construction CSCL scripts. This approach may be considered as advantageous in contrast to prior training or moderation of collaborative knowledge construction in CSCL environments, because learners may not be able to gather prior to CSCL in order to learn how to cooperate. Prior training is costly and moderation of CSCL is a particularly difficult enterprise, which may need additional qualification for teachers (Clark et al., 2003). CSCL scripts in contrast may warrant the quality of collaborative knowledge construction independent of the variable competencies of learners and teachers. CSCL scripts can have immediate effects on processes of knowledge co-construction. Prompts can support learners to construct relations between conceptual and problem space and prompts can encourage learners to disagree and inquire multiple perspectives (Nussbaum et al., 2002).

Chapter 8: Discussion

The results of this study suggest that especially *social cooperation scripts* should be considered in order to improve knowledge co-construction. Unguided interaction appears to be sub-optimal to some extent, but social cooperation scripts can help learners to interact in specific ways. This support, which is oriented towards social processes, appears to facilitate specific cognitive processes and outcomes.

The study also shows, however, that *epistemic cooperation scripts* can be detrimental for collaborative knowledge construction. Epistemic co-operation scripts can overprompt learners to some extent (Rosenshine et al., 1996). Important knowledge construction activities can be obstructed if scripts facilitate epistemic activities of learners on a very concrete and detailed level (Cohen, 1994). The epistemic cooperation script that has been examined in this study has suggested specific relations between conceptual and problem space. This script may substitute some cognitive processes that are important for knowledge construction. It has been shown that epistemic cooperation scripts can also be detrimental to social processes of collaborative knowledge construction. Some learners appear to focus on epistemic activities at the expense of some social processes (Dansereau, 1988). Therefore, instructional designers may need to carefully ponder whether they want to facilitate *effects with* or *effects of* the supported learning environment (Kollar et al., 2006; Salomon, 1993a).

Instructional support of epistemic activities needs to be designed carefully. Epistemic scripts should not support learners on a concrete level, but rather facilitate activities that support meta-cognition, e.g., question asking. In order to exploit the advantages of collaborative knowledge construction, instructional support needs to encourage learners to construct knowledge actively, whereas support that provides learners with an adequate solution may obstruct knowledge construction. In this way, epistemic cooperation scripts may be improved when meta-cognitive aspects are being considered (cf. King, 1999).

Recent studies emphasized the importance of *argumentation* for collaborative knowledge construction (e.g., Chinn et al., 2000; Keefer et al., 2000; Leitão, 2000). Argumentation structures in discourse may correspond with the reasoning of learners. These argumentation structures may, for in-

203

stance, include the presentation of various positions, backing of these positions with evidence or construction of counter-arguments. Little, however, is known with respect to how argumentative knowledge construction can be facilitated (Stegmann, Weinberger, & Fischer, 2007; Weinberger, Stegmann, Fischer, & Mandl, 2003). The facilitation of argumentation with cooperation scripts may be an approach which does not overscript learners in the sense of simplifying the task, but that challenges learners to monitor and improve their contributions. It has been argued that with the facilitation of learners' argumentation, one basic problem of collaborative knowledge construction can be solved. Learners may not actually construct knowledge together, but rather regress in discourse, because they lack the adequate task strategies and knowledge necessary to solve problems (cf. Tudge, 1989). Schwarz et al. (2000) found that even "two wrongs may make a right if they argue together." With the help of argumentative skills, learners may successfully challenge invalid ideas, which emerge in discourse. Learners can thereby identify adequate perspectives, which may foster knowledge acquisition. It is possible that argumentation can also be fostered with CSCL scripts (Stegmann et al., 2007). Argumentative cooperation scripts may reduce overscripting effects and facilitate learners to better construct knowledge. This may be particularly relevant, because even the social script, which lived up to expectations, had some detrimental effects on argumentation as the qualitative case studies show. Critics may be less concerned with backing up their critique by authority of proof when a script is providing a social kind of authority for critical statements. Studies on effects of social scripts for CSCL on argumentation appear to back up this assumption (Weinberger et al., 2003).

## 8.8    Limitations of the Study

The results of the study may need to be put into perspective with respect to their ecological validity, the analysis of collaborative knowledge construction processes, and the instructional approach of cooperation scripts.

The study has been conducted in a generally ecologically valid setting. Learners have experienced the CSCL session as part of the standard curriculum as the CSCL session substituted one session of a face-to-face seminar. On the grounds of prior findings and a constructivist perspective, it could be argued that the learners have constructed knowledge in a more active way than in the standard lecture or seminar, in which learners may not have the opportunity to produce 6 words per minute or more (Kern, 1995). The downturn of this highly ecological, but experimental setting is, however, that it is also highly specific. It is yet unclear to what extent less complex and less debatable problems may initiate discussion in similar ways. Therefore, the results need to be confirmed in a variety of settings with varying learning materials.

Furthermore, the study has regarded ad-hoc groups as one possible case of how learners co-construct knowledge in CSCL environments. Typically, learners in CSCL environments are initially anonymous, and some CSCL environments may foresee collaborative knowledge construction for groups which 'meet' online by chance and only for short periods of time. There is need, however, for empirical studies regarding the facilitation of groups who learn together online for a longer period. Further studies need to examine the effects of scripts for this long term collaboration. It may be particularly interesting to examine how discourse shifts from initially anonymous online groups to virtual communities, which accumulate a shared knowledge base. As scripts may substitute knowledge construction to some extent, the *fading* of instructional support such as scripts may be an important criterion for the facilitation of long-term collaborators (Collins et al., 1989).

One particular problem of CSCL and its facilitation is the varying degree of acceptance of computer-mediated learning (cf. Richter et al., 2001). Although learners may have perceived the learning environment of the study as part of their curriculum, it is yet unclear to what extent the CSCL environment with the specific prompt-based cooperation scripts would have been accessed when learners were not invited to a computer laboratory at a specified time. Short interviews in the debriefing phase did not reveal apparent differences regarding acceptance between the experimental conditions. Other studies indicate that learners may try to avoid participation in CMC, but may prefer to also meet FTF outside of a virtual environment (Hesse & Giovis, 1997; Reinmann-Rothmeier & Mandl, 2002).

The analysis of the various phenomena and effects of collaborative knowledge construction has been subject to some paradigmatic changes (Bruhn, 2000; Dillenbourg et al., 1995). After comparing collaborative to individual learning, research aimed to identify conditions relevant for collaborative knowledge construction (Slavin, 1993). An immense number of conditions and complex interactions have been identified (Renkl & Mandl, 1995), and had Mandl and Renkl (1992) 'plea for more local theories of collaborative knowledge construction' (see also Renkl, 1997). Analysis of processes has been regarded to as a way out of the complexity-dilemma of collaborative knowledge construction (Bruhn, 2000). There are, however, many approaches towards relevant co-construction processes (see for example Fischer, 2002). In the present study, the two most influential approaches have been analyzed, which build on the idea that discourse between learners is somehow related to the learning processes. Variance in learning outcome, however, may not yet be sufficiently explained for several reasons.

*First* of all, there may be explanations other than social and cognitive processes that account for the variance within collaborative knowledge construction. O'Donnell and Dansereau (1992), for instance, suggest that apart from social and cognitive aspects, affective and meta-cognitive dimensions may need to be analyzed and facilitated.

*Second*, these process dimensions may interact. In this respect, it was also necessary to analyze interaction of process with contextual variables. More variance may be clarified, if learning prerequisites entered the actual analysis. In this study, socio-cognitive approaches, which concentrate on specific processes of collaborative knowledge construction, have been outlined and utilized as tools to analyze and facilitate collaborative knowledge construction. Obviously, many more approaches may need to be applied to better understand collaborative knowledge construction. Educational psychology appears to not yet dispose of the one theoretical approach that sufficiently describes and predicts collaborative knowledge construction. Any study of collaborative knowledge construction may thus take only one or a limited number of perspectives towards the complex research object.

*Third*, research on collaborative knowledge construction is also limited with regard to the methodological repertoire. Apart from counting individual discourse phenomena, research has shifted to analyze sequences of discourse activities, because individual phenomena can take different meanings for collaborative knowledge construction, depending on discourse context. Furthermore, some theoretical approaches indicate that discourse structure is better at predicting learning outcome of collaborative knowledge construction than individual phenomena. For instance, transactivity may be better analyzed, not by identifying individual more or less transactive discourse activities, but by qualitatively analyzing structures of mutual referencing in discourse of learners. Neuman et al. (2000) applied a sequential analysis which detects whether the probability of specific discourse activities following specific other discourse activities differs from chance. Thus, they have found, for instance, that good problem solvers have regulations, i.e. plans to conduct specific activities like "I will do it in several steps," being followed by justifications for these regulations. Janetzko and Fischer (2002) propose a more advanced computer-supported approach which may identify more complex recurring patterns that are determined a priori by the researcher, supposedly based on theoretical assumptions. In this study, social modes and epistemic activities have been assessed in reference to discourse context (cf. Chi, 1997). The individual categories therefore give a description of relations between individual utterances. The social modes of co-construction in particular, have not been assessed as isolated units based on surface criteria like "word x = category x," but are interpretations of the individual function of the respective social mode in discourse. It is possible that the refinement of process measures is not as much a question of statistical method, but of granularity of the unit of analysis (cf. Chi, 1997). In order to replicate quantitative findings on processes of collaborative knowledge construction, qualitative analyses may be particularly useful. Therefore, additional graphical coding analyses have been presented with respect to case studies of each of the experimental conditions, which may give better impressions of collaborative knowledge construction processes when applied in correspondence with quantitative analyses.

Evidently, as collaborative knowledge construction is a complex research matter, the pivotal criterion is the correspondence of the theoretical and methodological repertoires. As processes of collaborative knowledge construction can be complex, they have been referred to as a new 'black box' by some researchers (Dillenbourg, 1999; Hogan et al., 2000). However, this 'black box' of collaboration can directly be subject to analysis and facilitation. Discourse in CSCL in particular may be assessable with more ease and can also be influenced more directly than was possible in FTF settings. In this respect, research on CSCL may not only clarify processes in a computer-supported setting, but also help to guide research on collaborative knowledge construction in general towards relevant phenomena and functions.

Apart from analyzing, the study aimed to facilitate processes and outcomes of collaborative knowledge construction. Prompt-based cooperation scripts appear to be a feasible and effective instructional approach for CSCL. A more fundamental limitation of this instructional approach, however, needs to be discussed. CSCL scripts have been applied as alternative instructional approach, e.g., to training learners to collaborate (Rummel & Spada, 2005), which may be difficult and costly to realize in CSCL environments. The rationale of this instructional approach is that specific discourse activities are suggested, and thus, learning outcome is also facilitated. As the results show, this instructional support may overscript learners, however, in the sense that they do not internalize the suggested task strategies to apply knowledge to problem case information. This *failure of internalization* could possibly indicate a more general problem of instructional support that substitutes what it aims to foster (Reiser, 2002). In this way, the social cooperation script may have impeded the acquisition of cooperative strategies. Learners of groups without social cooperation script may have had the chance to consider better ways of collaboration for the next time they are supposed to collaborate in learning groups. The social script learners, however, may have not internalized the collaboration strategies the social script suggested. Even though there is no data supporting this assumption, the instructional approach builds on the idea that learners may not need to acquire collaborative or task strategies when they are supported with the respective scripts. Given that learners recognize a script as conveying a specific beneficial strategy interesting to the learners, scripts may also model behavior, which learners could eventually imitate. To successfully model behavior through scripts may largely depend on the motivational structure of the learning environment.

Prompt-based cooperation scripts may be applied with care for learning groups who have not yet developed a self-supporting collaborative learning culture. In order to initiate sustainable development, CSCL scripts should not substitute, but support already existing, adequate scripts sensu Schank and Abelson (1977) for collaborative knowledge construction. For a time of transition towards self-sustained collaborative learning cultures, CSCL scripts may, however, improve acceptance and efficiency of collaborative knowledge construction (see Renkl et al., 1996). In this respect it is important to note that the CSCL scripts conceptualized in this work may have given very detailed instructions which were argued to be detrimental for complex learning tasks, but also provided learners with sufficient degrees of freedom regarding whether the detailed instructions must actually be used or could be ignored (Cohen, 1994; Dillenbourg, 2002; Veerman & Treasure-Jones, 1999).

When summarizing the limitations of the study, several gaps can be identified. The first gap can be noted between the analysis of processes of collaborative knowledge construction and the development of instructional support. It may be pointless to design scripts that regard all social modes or epistemic activities, for instance. Instead, externally induced cooperation scripts may need to be adopted to internally represented scripts. That means, that scripts should only facilitate those activities of collaborative knowledge construction that learners do not spontaneously engage in. It may be particularly difficult, however, to identify the individual deficits of learners regarding scripts. The starting point of the present study was, for instance, that learners engaged in satisficing regarding specific social and cognitive processes. Thus, elicitation and conflict-oriented consensus building have been facilitated in the present study. Future technology may automatically assess learners' deficits while collaboratively constructing knowledge and suggest or automatically implement specific scripts or script components (Dönmez, Rosé, Stegmann, Weinberger, & Fischer, 2005; Kobbe, Weinberger, Dillenbourg, Harrer, Hämäläinen, Häkkinen, & Fischer, 2007). A second gap can be identified between the aims of the instructional support and the actual effects of the instructional support on processes of collaborative knowledge construction. For instance, the design of the present study did not foresee that learners facilitated with the epistemic script would be impeded in elicitation. Scripts may have 'side effects' or foster collaborative knowledge construction only at certain costs. Lastly, there may be a gap between what can be observed in discourse and internal processes of knowledge construction. In order to examine this distortion between discourse phenomena and learning, more case studies may need to be applied in order to trace specific (mis-)conceptions of learners throughout their collaborative phase and their individual post-test activities. Furthermore, participants may need to be interviewed in detail in order to better understand what can be observed in discourse. In-depth interviews may not only aim to examine cognitive structures, but also to reveal goal orientations, interests, and epistemological beliefs of learners regarding the subject matter.

## 8.9 Conclusion

The most fundamental, recent change to learning has been motivated by the introduction of computers as learning tools. This impact on learning has been somewhat limited thus far improving access to education, rather than improving quality of education. Beyond the practical advantages of making education more accessible, text-based CMC, in particular, has been argued to pose potentials for collaborative knowledge construction. Learners may "see what they built together" (Pea, 1994), because the discourse processes are being recorded on a central database. Learners have more time to formulate their contributions. They may therefore reflect better on what they are going to contribute and what their learning partners have put forward (Cohen & Scardamalia, 1998; Mason, 1998). Apparently text-based CMC is facilitating epistemic discourse – learners concentrate on the learning task rather than to engage in non-epistemic activities (Woodruff, 1995). Furthermore, learners should participate actively in mutual construction of a shared conception of a problem and a strategy to solve it. Text-based CMC may reduce production-blocking effects and motivate learners to contribute (Kern, 1995). Learners in text-based CMC may then be judged by their contributions, rather than their non-salient social background. Therefore, learners who usually remain silent in class or whose active participation in learning groups is unlikely may profit from CSCL environments with text-based communication interfaces. However, there are indications that learners do not exploit these potentials of CSCL with text-based CMC, but rather reproduce sub-optimal behavior regarding learning known in FTF collaborative knowledge construction. For instance, learners rarely realize the chance for equal and active participation in text-based CMC, but instead use text-based CMC to "lurk" onto group processes (Hesse & Giovis, 1997; Weinberger & Mandl, 2001). Learners may lack adequate scripts sensu Schank and Abelson (1977) to collaboratively construct knowledge in CSCL environments. Therefore, CSCL may need instructional support that builds on acknowledged approaches of educational psychology. The approach analyzed in this work is to externally induce CSCL scripts. Ultimately, learners need to be facilitated to internalize functional scripts themselves. In order to continuously conceptualize better CSCL scripts, existing internal scripts may need to be identified and endorsed with specifically designed externally induced scripts. In conclusion, learners should be facilitated to flexibly apply a variety of scripts and to internalize the underlying principles of externally induced scripts.

The results of the study suggest that this instructional support needs to be carefully designed in order to maintain the complexity of tasks for collaborative knowledge construction and to support a specific social structure of discourse (Palincsar & Herrenkohl, 1999; Reiser, 2002; Wiley & Voss, 1999). Future studies of CSCL may further investigate what aspects of collaborative knowledge construction instructional support should aim at, e.g., argumentative knowledge construction, and how instructional support may be attuned to avoid overscripting. These studies should be theory-driven rather than guided solely by what is technologically feasible.

Future studies may not only respect complexity of collaborative knowledge construction with regard to conceptualizing instructional support, but may also refer to multiple theoretical approaches and research traditions when analyzing collaborative knowledge construction.

Educational Psychology may further refine approaches to analyze and facilitate specific processes and outcomes of collaborative knowledge construction. CSCL research may provide technically advanced forms to implement instructional approaches and reduce deficits of text-based CMC regarding coordination of learners. Blending both research traditions, learning environments may be conceptualized, realized, and actually implemented into educational practice to qualitatively change learning through CSCL.

# 9 References

Anderson, A. H., Smallwood, L., MacDonald, R., Mullin, J., Fleming, A., & O'Malley, C. (2000). Video data and video links in mediated communication: What do users value? *International Journal of Human-Computer Studies, 52*(1), 165-187.

Anderson, R. C., Chinn, C., Chang, J., Waggoner, M., & Yi, H. (1997). On the logical integrity of children's arguments. *Cognition and Instruction, 15*(2), 153-167.

Aronson, E., Blaney, N., Stephan, G., Silkes, J., & Snapp, M. (1978). *The jigsaw classroom.* Beverly Hills, CA: Sage.

Baker, M., & Lund, K. (1997). Promoting reflective interactions in a CSCL environment. *Journal of Computer Assisted Learning, 13*, 175-193.

Bandura, A. (1986). *Social foundations of thought and action.* Englewood Cliffs, NJ: Prentice-Hall.

Barrows, H. S., & Tamblyn, R. (1980). *Problem-based learning.* New York: Springer.

Berkowitz, M. W., & Gibbs, J. C. (1983). Measuring the development of features of moral discussion. *Merrill-Palmer Quarterly, 29*, 399-410.

Boring, E. G. (1923). Intelligence as the tests test it. *New Republic, 35*, 35-37.

Boshuizen, H. P. A., & Schmidt, H. G. (1992). On the role of biomedical knowledge in clinical reasoning by experts, intermediates and novices. *Cognitive Science, 16*, 153-184.

Bransford, J. D., Franks, J. J., Vye, N. J., & Sherwood, R. D. (1989). New approaches to learning and instruction: Because wisdom can't be told. In S. Vosniadou & A. Ortony (Eds.), *Similarity and analogical reasoning* (pp. 470-497). Cambridge: Cambridge University Press.

Bromme, R., Hesse, F. W., & Spada, H. (Eds.). (2005). *Barriers and biases in computer-mediated knowledge communication - and how they may be overcome.* Boston: Kluwer.

Bromme, R., & Jucks, R. (2001). Wissensdivergenz und Kommunikation: Lernen zwischen Experten und Laien im Netz [Knowledge divergence and communication: Online learning between experts and laypersons]. In F. W. Hesse & H. F. Friedrich (Eds.), *Partizipation und Interaktion im virtuellen Seminar [Participation and interaction in the virtual seminar]* (pp. 81-103). Münster: Waxmann.

Brown, A. L., & Palincsar, A. S. (1989). Guided, cooperative learning and individual knowledge acquisition. In L. B. Resnick (Ed.), *Knowing, learning, and instruction. Essays in the honour of Robert Glaser* (pp. 393-451). Hillsdale, NJ: Erlbaum.

Bruhn, J. (2000). *Förderung des kooperativen Lernens über Computernetze [Facilitation of cooperative learning via computer networks].* Frankfurt a. M.: Lang.

Bruhn, J., Gräsel, C., Fischer, F., & Mandl, H. (1997). *Kategoriensystem zur Erfassung der Kokonstruktion von Wissen im Diskurs [Coding system for analysis of knowledge co-construction].* Unpublished manuscript, Ludwig Maximilian University of Munich, Institute of Educational Psychology.

Cannon-Bowers, J. A., & Salas, E. (1998). Team Performance and Training in Complex Environments: Recent Findings From Applied Research. *Current Directions In Psychological Science, 7*(3), 83-87.

Cannon-Bowers, J. A., & Salas, E. (2001). Reflections on shared cognition. *Journal of Organizational Behavior, 22*, 195-202.

Carletta, J., Anderson, A. H., & McEwan, R. (2000). The effects of multimedia communication technology on non-collocated teams: A case study. *Ergonomics, 43*(8), 1237-1251.

Chan, C. K. K. (2001). Peer collaboration and discourse patterns in learning from incompatible information. *Instructional Science, 29*, 443-479.

Chan, C. K. K., Burtis, P. J., & Bereiter, C. (1997). Knowledge building as a mediator of conflict in conceptual change. *Cognition and Instruction, 15*(1), 1-40.

Chi, M. T. H. (1997). Quantifying qualitative analyses of verbal data: A practical guide. *Journal of the Learning Sciences, 6*, 271-315.

Chi, M. T. H., & Bassok, M. (1989). Learning from examples via self-explanations. In L. B. Resnick (Ed.), *Knowing, learning and instructions. Essays in the honor of Robert Glaser* (pp. 251-282). Hillsdale, NJ: Erlbaum.

Chi, M. T. H., De Leeuw, N., Chiu, M. H., & LaVancher, C. (1994). Eliciting self-explanations improves understanding. *Cognitive Science, 18*, 439-477.

Chi, M. T. H., Feltovich, P. J., & Glaser, R. (1981). Categorization and representation of physics problems by experts and novices. *Cognitive Science, 5*, 121-152.

Chinn, C. A., & Brewer, W. F. (1993). The role of anomalous data in knowledge aquisition: A theoretical framework and implications for science instruction. *Review of Educational Research, 63*, 1-49.

Chinn, C. A., O'Donnell, A. M., & Jinks, T. S. (2000). The structure of discourse in collaborative learning. *The Journal of Experimental Education, 69*(1), 77-97.

Clark, D., Weinberger, A., Jucks, R., Spitulnik, M., & Wallace, R. (2003). Designing effective science inquiry in text-based computer supported collaborative learning environments. *International Journal of Educational Policy, Research & Practice, 4*(1), 55-82.

Clark, H. H. (1992). *Arenas of language use.* Chicago: University of Chicago Press.

Clark, H. H., & Brennan, S. E. (1991). Grounding in communication. In S. D. Teasley (Ed.), *Perspectives on socially shared cognition* (pp. 127-149). Washington, DC: American Psychologist Association.

Clark, H. H., & Marshall, C. R. (1981). Definite references and mutual knowledge. In A. K. Joshi, B. L. Webber & I. A. Sag (Eds.), *Elements of discourse understanding* (pp. 10-63). Cambridge: Cambridge University Press.

Clark, H. H., & Schaefer, E. F. (1989). Contributing to discourse. *Cognitive Science, 13*, 259-294.

Clark, H. H., & Wilkes-Gibbs, D. (1986). Referring as a collaborative process. *Cognition, 22*, 1-39.

Clark, R. E. (1994). Media will never influence learning. *Educational Technology Research and Development, 42*(2), 21-29.

Cobb, P. (1988). The tensions between theories of learning and instruction in mathematics education. *Educational Psychologist, 23*, 78-103.

Cobb, P., & Bowers, J. (1999). Cognitive and situated learning. Perspectives in theory and practice. *Educational Researcher, 28*(2), 4-15.

Cognition and Technology Group at Vanderbilt. (1992). The Jasper series as an example of anchored instruction: Theory, program, description, and assessment data. *Educational Psychologist, 27*, 291-315.

Cognition and Technology Group at Vanderbilt. (1997). *The Jasper Project: Lessons in curriculum, instruction, assessment, and professional development.* Mahwah, NJ: Erlbaum.

Cohen, A., & Scardamalia, M. (1998). Discourse about ideas: Monitoring and regulation in face-to-face and computer-mediated environments. *Interactive Learning Environments, 6*(1-2), 93-113.

Cohen, E. G. (1994). Restructuring the classroom: Conditions for productive small groups. *Review of Educational Research, 64*, 1-35.

Cohen, E. G., & Lotan, R. A. (1995). Producing equal-status interaction in the heterogeneous classroom. *American Educational Research Journal, 32*, 99-120.

Cohen, J. (1986). Theoretical consideration of peer tutoring. *Psychology in the Schools, 23*(2), 175-186.

Coleman, E. B. (1998). Using explanatory knowledge during collaborative problem solving in science. *The Journal of the Learning Sciences, 7*(3&4), 387-427.

Collins, A., & Bielayczyc, K. (1997). *Dreams of technology-supported learning communities.* Paper presented at the Sixth International Conference on Computer-Assisted Instruction, Taipeh, Taiwan.

Collins, C., Brown, J. S., & Newmann, S. E. (1989). Cognitive apprenticeship: Teaching the crafts of reading, writing and mathematics. In L. B. Resnick (Ed.), *Knowing, learning, and instruction: Essays in honor of Robert Glaser* (pp. 453-494). Hillsdale, NJ: Erlbaum.

Cooke, N. J., Salas, E., Cannon-Bowers, J. A., & Stout, R. (2000). Measuring team knowledge. *Human Factors, 42*, 151-173.

Crook, C. (1995). On resourcing a concern for collaboration with peer interactions. *Cognition and Instruction, 13*(4), 541-547.

Dalbert, C. (1996). Ungewißheitstoleranz und der Umgang mit Ungerechtigkeit [Ambiguity tolerance and dealing with injustice]. In C. Dalbert (Ed.), *Über den Umgang mit Ungerechtigkeit [Dealing with Injustice]* (pp. 189-230). Bern: Huber.

Dansereau, D. F. (1988). Cooperative learning strategies. In C. E. Weinstein, E. T. Goetz & P. A. Alexander (Eds.), *Learning and study strategies: Issues in assessment, instruction, and evaluation* (pp. 103-120). Orlando, FL: Academic Press.

Dansereau, D. F., Collins, K. W., McDonald, B. A., Holley, C., Garland, J., Diekhoff, G., & Evans, S. H. (1979). Development and evaluation of a learning strategy training programm. *Journal of Educational Psychology, 71*, 64 - 73.

De Grave, W. S., Boshuizen, H. P. A., & Schmidt, H. G. (1996). Problem based learning: Cognitive and metacognitive processes during problem analysis. *Instructional Science, 24*, 321-341.

De Jong, T., & Fergusson-Hessler, M. G. M. (1996). Types and qualities of knowledge. *Educational Psychologist, 31*, 105-113.

De Jong, T., & van Joolingen, W. (1998). Scientific discovery learning with computer simulations of conceptual domains. *Review of Educational Research, 68*(2), 179-201.

De Lisi, R., & Goldbeck, S. L. (1999). Implication of Piagetian theory for peer learning. In A. M. O'Donnell & A. King (Eds.), *Cognitive perspectives on peer learning* (pp. 3-37). Mahwah, NJ: Erlbaum.

Derry, S. J. (1999). A fish called peer learning: Searching for common themes. In A. M. O'Donnell & A. King (Eds.), *Cognitive perspectives on peer learning* (pp. 197-211). Mahwah, NJ: Erlbaum.

Dewey, J., & Bentley, A. F. (1949). *Knowing and the known.* Boston: Beacon Press.

Diehl, M., & Ziegler, R. (2000). Informationsaustausch und Ideensammlung in Gruppen [Information exchange and brainstorming in groups]. In M. Boos, K. J. Jonas & K. Sassenberg (Eds.), *Computervermittelte Kommunikation in Organisationen [Computer-mediated communication in organizations]* (pp. 89-103). Göttingen: Hogrefe.

Dillenbourg, P. (1999). Introduction: What do you mean by "collaborative learning"? In P. Dillenbourg (Ed.), *Collaborative Learning. Cognitive and computational approaches* (pp. 1-19). Amsterdam: Pergamon.

Dillenbourg, P. (2002). Over-scripting CSCL: The risks of blending collaborative learning with instructional design. In P. A. Kirschner (Ed.), *Three worlds of CSCL: Can we support CSCL?* (pp. 61-91). Heerlen: Open Universiteit Nederland.

Dillenbourg, P., Baker, M., Blaye, A., & O'Malley, C. (1995). The evolution of research on collaborative learning. In P. Reimann & H. Spada (Eds.), *Learning in humans and machines: Towards an interdiciplinary learning science* (pp. 189-211). Oxford: Elsevier.

Ditton, H. (1998). *Mehrebenenanalyse. Grundlagen und Anwendungen des Hierarchisch Linearen Modells [Multi-level analysis. Foundations and applications of the hierarchical linear model].* Weinheim: Juventa.

Dochy, F. J. R. C. (1992). *Assessment of prior knowledge as a determinant for future learning.* Utrecht, NL: Lemma.

Dochy, F. J. R. C., Segers, M., van den Bossche, P., & Gijbels, D. (2003). Effects of problem-based learning: A meta-analysis. *Learning and Instruction, 13*(5), 533-568.

Doise, W., & Mugny, G. (1984). *The Social Development of the Intellect.* Oxford: Pergamon.

Dönmez, P., Rosé, C. P., Stegmann, K., Weinberger, A. & Fischer, F. (2005). Supporting CSCL with automatic corpus analysis technology. In T. Koschmann, D. Suthers, & T. W. Chan (Eds.), *Proceedings of the International Conference on Computer Supported Collaborative Learning - CSCL 2005* (pp. 125-134). Taipei: Erlbaum.

Döring, N. (1997a). Kommunikation im Internet: Neun theoretische Ansätze [Internet communication: Nine theoretical approaches]. In B.

Batinic (Ed.), *Internet für Psychologen [Internet for psychologists]* (pp. 267-289). Göttingen: Verlag für Psychologie.

Döring, N. (1997b). Lernen und Lehren im Internet [Learning and teaching on the internet]. In B. Batinic (Ed.), *Internet für Psychologen [Internet for psychologists]* (pp. 359-393). Göttingen: Verlag für Psychologie.

Döring, N. (1999). *Sozialpsychologie des Internets. Die Bedeutung des Internets für Kommunikationsprozesse, Identitäten, soziale Beziehungen und Gruppen. [The social psychology of the internet. The importance of the internet for communication processes, identities, social relations and groups].* Göttingen: Hogrefe.

Dubrovsky, V. J., Kiesler, S., & Sethna, B. N. (1991). The equalization phenomenon: Status effects in computer-mediated and face-to-face decision-making groups. *Human-Computer Interaction, 6,* 119-146.

Dufresne, R. J., Gerace, W. J., Thibodeau Hardiman, P., & Mestre, J. P. (1992). Constraining novices to perform expertlike problem analyses: Effects on schema acquisition. *The Journal of the Learning Sciences, 2*(3), 307-331.

Fabos, B., & Young, M. D. (1999). Telecommunication in the classroom: Rhetoric versus reality. *Review of Educational Research, 69*(3), 217-259.

Fischer, F. (1998). *Mappingverfahren als kognitive Werkzeuge für problemorientiertes Lernen [Mapping procedures as cognitive tools for problem-oriented learning].* Frankfurt a. M.: Lang.

Fischer, F. (2001). *Gemeinsame Wissenskonstruktion. Analyse und Förderung in computerunterstützten Kooperationsszenarien [Collaborative knowledge construction. Analysis and facilitation in computer-supported collaborative scenarios].* Unpublished professorial dissertation, Ludwig Maximilian University of Munich.

Fischer, F. (2002). Gemeinsame Wissenskonstruktion - theoretische und methodologische Aspekte [Collaborative knowledge construction - theoretical and methodological aspects]. *Psychologische Rundschau, 53*(3), 119-134.

Fischer, F., Bruhn, J., Gräsel, C., & Mandl, H. (2000). Kooperatives Lernen mit Videokonferenzen: Gemeinsame Wissenskonstruktion und individueller Lernerfolg [Cooperative learning through video conferencing: Collaborative knowledge construction and individual learning success]. *Kognitionswissenschaft, 9*(1), 5-16.

Fischer, F., Bruhn, J., Gräsel, C., & Mandl, H. (2002). Fostering collaborative knowledge construction with visualization tools. *Learning and Instruction, 12*, 213-232.

Fischer, F., Gräsel, C., Kittel, A., & Mandl, H. (1996). Entwicklung und Untersuchung eines computerbasierten Mappingverfahren zur Strukturierung komplexer Information [Development and analysis of a computer-based mapping tool for structuring complex information]. *Psychologie in Erziehung und Unterricht, 43*, 266-280.

Fischer, F., & Mandl, H. (2000). Förderung der Strategieanwendung mit Expertenmaps [Fostering strategy application through the use of expert maps]. In D. Leutner & R. Brünken (Eds.), *Neue Medien in Unterricht, Aus- und Weiterbildung. Aktuelle Ergebnisse empirischer pädagogischer Forschung [New media in instruction, education, and professional development. Current findings of empirical pedagogical research]* (pp. 47-56). Münster: Waxmann.

Fischer, F., & Mandl, H. (2001). Facilitating the construction of shared knowledge with graphical representation tools in face-to-face and computer-mediated scenarios. In P. Dillenbourg & A. Eurelings & K. Hakkarainen (Eds.), *European perspectives on computer-supported collaborative learning* (pp. 230-236). Maastricht, NL: University of Maastricht.

Fischer, F., & Mandl, H. (2002). Lehren und Lernen mit neuen Medien [Teaching and learning with new media]. In R. Tippelt (Ed.),

*Handbuch der Bildungsforschung [Handbook of educational research]* (pp. 627-641). Opladen: Leske & Budrich.

Fischer, F., Mandl, H., Haake, J., & Kollar, I. (Eds.). (2007). *Scripting computer-supported communication of knowledge - cognitive, computational and educational perspective.* New York: Springer.

Gagné, R. M. (Ed.). (1987). *Instructional Technology: Foundations.* Hillsdale, NJ: Erlbaum.

Ge, X., & Land, S. M. (2002, April). *The effects of question prompts and peer interactions in scaffolding students' problem-solving processes on an ill-structured task.* Paper presented at the Annual Meeting of the American Educational Research Association, New Orleans, LA.

Gelman, R., & Greeno, J. G. (1989). On the nature of competence: Principles for understanding in a domain. In L. B. Resnick (Ed.), *Knowing, learning, and instruction: Essays in honor of Robert Glaser* (pp. 125-187). Hillsdale, NJ: Erlbaum.

Gerstenmaier, J., & Mandl, H. (1999). Konstruktivistische Ansätze in der Erwachsenenbildung und Weiterbildung [Constructivist approaches in adult education and professional development]. In R. Tippelt (Ed.), *Handbuch Erwachsenenbildung / Weiterbildung [Handbook of adult education / professional development]* (pp. 184-192). Opladen: Leske & Buldrich.

Gerstenmaier, J., & Mandl, H. (2001). Methodologie und Empirie zum Situierten Lernen [Methodology and empirics of situated learning]. *Schweizerische Zeitschrift für Bildungswissenschaften, 3*(23), 453-470.

Glaser, R. (1991). The maturing of the relationship between the science of learning and cognition and educational practice. *Learning and Instruction, 1*, 129-144.

Gordin, D. N., Gomez, L. S., Pea, R. D., & Fishman, B. J. (1996). Using the World Wide Web to build learning communities in K-12. *Journal of Computer-Mediated Communication, 2*(3).

Gräsel, C. (1997). *Problemorientiertes Lernen [Problem-oriented learning]*. Göttingen: Hogrefe.

Gräsel, C., & Fischer, F. (2000). Information and communication technologies at schools. A trigger for better teaching and learning? *International Journal of Educational Policy, Research and Practice, 1*(3), 327-336.

Gräsel, C., Fischer, F., Bruhn, J., & Mandl, H. (2001). "Let me tell you something you do know". In S. Dijkstra, D. Jonassen & D. Sembill (Eds.), *Multimedia Learning. Results and Perspectives* (pp. 111-137). Frankfurt: Lang.

Gräsel, C., Fischer, F., & Mandl, H. (2001). The use of additional information in problem-oriented learning environments. *Learning Environments Research, 3*, 287-305.

Gräsel, C., & Mandl, H. (1993). Förderung des Erwerbs diagnostischer Strategien in fallbasierten Lernumgebungen [Facilitating the acquisition of diagnostic strategies in case-based learning environments]. *Unterrichtswissenschaft, 21*, 355-370.

Gräsel, C., Mandl, H., Fischer, M., & Gärtner, R. (1994). Vergebliche Designermüh? Interaktionsangebote in problemorientierten Computerlernprogrammen [Futile designer efforts? Interactive elements in computer-based learning environments]. *Unterrichtswissenschaft, 22*, 312-333.

Greeno, J. G. (1998). The situativity of knowing, learning, and research. *American Psychologist, 53*, 5-26.

Gruber, H., Renkl, A., & Schneider, W. (1994). Expertise and memory development: Longitudinal findings from the domain of chess. *The German Journal of Psychology, 18*, 243-244.

Hakkarainen, K., & Palonen, T. (2003). Patterns of female and male students' participation in peer interaction in computer-supported learning. *Computers & Education, 40*, 327-342.

Hatano, G., & Inagaki, K. (1991). Sharing cognition through collective comprehension activity. In L. B. Resnick, J. M. Levine & S. D. Teasley (Eds.), *Perspectives on socially shared cognition* (pp. 331-348). Washington, DC: American Psychological Association.

Heider, F. (1958). *The Psychology of Interpersonal Relations*. New York: Wiley.

Herrenkohl, L. R., & Guerra, M. R. (1998). Participant structures, scientific discourse, and student engagement in fourth grade. *Cognition and Instruction, 16*, 433-475.

Hertz-Lazarowitz, R., Benveniste Kirkus, V., & Miller, N. (1992). Implications of current research on cooperative interaction for classroom application. In R. Hertz-Lazarowitz & N. Miller (Eds.), *Interaction in cooperative groups. The theoretical anatomy of group learning* (pp. 253-280). Cambridge: Cambridge University Press.

Hesse, F. W., Garsoffky, B., & Hron, A. (1997). Interface-Design für computerunterstütztes kooperatives Lernen [Interface design for computer-supported cooperative learning]. In L. J. Issing & P. Klimsa (Eds.), *Information und Lernen mit Multimedia [Information and learning with multimedia]* (2nd ed., pp. 253-267). Weinheim: Beltz.

Hesse, F. W., & Giovis, C. (1997). Struktur und Verlauf aktiver und passiver Partizipation beim netzbasierten Lernen in virtuellen Seminaren [Structure and processes of active and passive participation in net-based learning through virtual seminars]. *Unterrichtswissenschaft, 25*(1), 34-55.

Hiltz, S. R., Johnson, K., & Turoff, M. (1986). Experiments in group decision making. Communication process and outcome in face-to-face versus computerized conferences. *Human Communication Research, 13*(2), 225-252.

Hofer, M., & Pikowsky, B. (1993). Validation of a category system of arguments in conflict discourse. *Argumentation, 7*, 135-148.

Hogan, K., Nastasi, B. K., & Pressley, M. (2000). Discourse patterns and collaborative scientific reasoning in peer and teacher-guided discussions. *Cognition and Instruction, 17*(4), 379-432.

Howe, C., & Tolmie, A. (1999). Productive interaction in the context of computer-supported collaborative learning in science. In K. Littleton & P. Light (Eds.), *Learning with computers: Analysing productive interaction* (pp. 24-45). London: Routledge.

Howe, C., Tolmie, A., & MacKenzie, M. (1995). Computer support for the collaborative learning of physics concepts. In C. O'Malley (Ed.), *Computer supported collaborative learning* (Vol. 128, pp. 303). Berlin: Springer.

Hron, A., Hesse, F. W., Cress, U., & Giovis, C. (2000). Implicit and explicit dialogue structuring in virtual learning groups. *British Journal of Educational Psychology, 70*, 53-64.

Hron, A., Hesse, F. W., Reinhard, P., & Picard, E. (1997). Strukturierte Kooperation beim computerunterstützten kollaborativen Lernen [Structured cooperation in computer-supported collaborative learning]. *Unterrichtswissenschaft, 25*(1), 56-69.

Huber, A. A. (1999). *Bedingungen effektiven Lernens in Kleingruppen unter besonderer Berücksichtigung der Rolle von Lernskripten [Conditions for effective learning of small groups with respect to the role of learning scripts]*. Schwangau: Huber.

Huber, G. L. (1987). Kooperatives Lernen: Theoretische und praktische Herausforgerung für die Pädagogische Psychologie [Cooperative learning: Theoretical and practical challenges for educational psychology]. *Zeitschrift für Entwicklungspsychologie und Pädagogische Psychologie, 19*(4), 340-362.

Huber, G. L., Sorrentino, R. M., Davidson, M. A., Eppler, R., & Roth, J. W. H. (1992). Uncertainty orientation and cooperative learning: Individual differences within and across cultures. *Learning and Individual Differences, 4*, 1-24.

Hytecker, V. I., Dansereau, D. F., & Rocklin, T. R. (1988). An analysis of the processes influencing the structured dyadic learning environment. *Educational Psychologist, 23*(1), 23-37.

Janetzko, D., & Fischer, F. (2002). Analyzing sequential data in computer-supported collaborative learning. In G. Stahl (Ed.), *Computer support for collaborative learning: Foundations for a CSCL community. Proceedings of the Conference on Computer Support for Collaborative Learning (CSCL) 2002 in Boulder, Colorado, USA* (pp. 585-586). Hillsdale, NJ: Erlbaum.

Johnson, D. W., & Johnson, R. T. (1992). Positive interdependence: Key to effective cooperation. In R. Hertz-Lazarowitz & N. Miller (Eds.), *Interaction in Cooperative Groups* (pp. 174-199). Cambridge: Cambridge University Press.

Jonassen, D. H., Campbell, J. P., & Davidson, M. E. (1994). Learning with media: Restructuring the debate. *Educational Technology Research and Development, 42*(2), 31-39.

Joseph, G. M., & Patel, V. L. (1990). Domain knowledge and hypothesis generation in diagnostic reasoning. *Medical Decision Making, 10*(1), 31-46.

Jucks, R., Bromme, R., & Runde, A. (2003). Audience Design von Experten in der netzgestützten Kommunikation: Die Rolle von Heuristiken über das geteilte Vorwissen [Experts' audience design in network-based communication: the role of heuristics over shared prior knowledge]. *Zeitschrift für Psychologie, 211*, 60-74.

Kaiser, F. J. (1983). *Theorie und Praxis der Fallstudiendidaktik [Theory and application of case-based didactics]*. Bad Heilbrunn: Klinkhardt.

Kassirer, J. P. (1995). Teaching problem-solving - how are we doing? *The New England Journal of Medicine, 332*, 1507-1509.

Keefer, M. W., Zeitz, C. M., & Resnick, L. B. (2000). Judging the quality of peer-led student dialogues. *Cognition and Instruction, 18*(1), 53-81.

Kern, R. G. (1995). Restructuring classroom interaction with networked computers: Effects on quantity and characteristics of language production. *The Modern Language Journal, 79*, 457-476.

Kerr, N. L., & Bruun, S. E. (1983). Dispensability of member-effort and group motivation loss: Free-rider effects. *Journal of Personality and Social Psychology, 44*, 78-94.

Kiesler, S. (1992). Talking, teaching, and learning in network groups: Lessons from research. In A. Kaye (Ed.), *Collaborative learning through computer conferencing. The Najaden Papers* (pp. 147-165). Berlin: Springer.

Kiesler, S., Siegel, J., & McGuire, T. W. (1984). Social psychological aspects of computer-mediated communication. *American Psychologist, 39*(10), 1123-1134.

Kiesler, S., & Sproull, L. (1992). Group decision making and communication technology. *Organizational Behavior and Human Decision Processes, 52*, 96-123.

King, A. (1989a). Effects of self-questioning training on college students' comprehension of lectures. *Contemporary Educational Psychology, 14*, 1-16.

King, A. (1989b). Verbal interaction and problem-solving within computer-assisted cooperative learning groups. *Contemporary Educational Psychology, 5*, 1-15.

King, A. (1990). Enhancing peer interaction and learning in the classroom through reciprocal questioning. *American Educational Research Journal, 27*, 664-687.

King, A. (1992). Faciliating elaborative learning through guided student-generated questioning. *Educational Psychologist, 27*, 111-126.

King, A. (1994). Guiding knowledge construction in the classroom: Effects of teaching children how to question and how to explain. *American Educational Research Journal, 31*, 338-368.

King, A. (1999). Discourse patterns for mediating peer learning. In A. M. O'Donnell & A. King (Eds.), *Cognitive perspectives on peer learning* (pp. 87-115). Mahwah, NJ: Erlbaum.

King, K. S. (1998). Designing 21st-century educational networlds: Structuring electronic social spaces. In C. J. Bonk & K. S. King (Eds.), *Electronic collaborators: Learner-centered technologies for literacy, apprenticeship, and discourse* (pp. 365-383). Mahwah, NJ: Erlbaum.

Kitchner, K. S. (1983). Cognition, metacognition, and epistemic cognition: A three-level model of cognitive processing. *Human Development, 26*, 222-232.

Kobbe, L, Weinberger, A., Dillenbourg, P., Harrer, A., Hämäläinen, R., Häkkinen, P., & Fischer, F. (2007). Specifying computer-supported collaboration scripts. *International Journal of Computer-Supported Collaborative Learning, 2*(2-3), 211-224.

Kollar, I., Fischer, F., & Hesse, F. (2006). Cooperation scripts - a conceptual analysis. *Educational Psychology Review, 18*, 159-185.

Koschmann, T. (1996). Paradigm shift and instructional technology: An introduction. In T. Koschmann (Ed.), *CSCL: Theory and practice of an emerging paradigm* (pp. 1-23). Mahwah, NJ: Erlbaum.

Koschmann, T., Kelson, A. C., Feltovich, P. J., & Barrows, H. S. (1996). Computer-supported problem-based learning: A principled approach to the use of computers in collaborative learning. In T. Koschmann (Ed.), *CSCL: Theory and practice of an emerging paradigm* (pp. 83-124). Mahwah, NJ: Erlbaum.

Krapp, A. (1999). Interest, motivation and learning: An educational-psychological perspective. *European Journal of Psychology in Education, 14*, 23-40.

Kruger, A. (1992). The effect of peer and adult-child transactive discussion on moral reasoning. *Merrill-Palmer Quarterly, 38*, 191-211.

Kruger, A., & Tomasello, M. (1986). Transactive discussion with peer and adults. *Developmental Psychology, 22*, 681-685.

Kuhn, D. (2001). How do people know? *Psychological Science, 12*, 1-8.

Lambiotte, J. G., Dansereau, D. F., O'Donnell, A. M., Young, M. D., Skaggs, L. P., & Hall, R. H. (1988). Effects of cooperative script manipulations on initial learning and transfer. *Cognition and Instruction, 5*(2), 103-121.

Lambiotte, J. G., Dansereau, D. F., O'Donnell, A. M., Young, M. D., Skaggs, L. P., Hall, R. H., & Rocklin, T. R. (1987). Manipulating cooperative scripts for teaching and learning. *Journal of Educational Psychology, 79*(4), 424-430.

Larkin, J. H., McDermott, J., Simon, D. P., & Simon, H. A. (1980). Expert and novice performance in solving physics problems. *Science, 208*, 1335-1342.

Larson, C. O., Dansereau, D. F., O'Donnell, A. M., Hytecker, V. I., Lambiotte, J. G., & Rocklin, T. R. (1985). Effects of metacognitive and elaborative activity on cooperative learning and transfer. *Contemporary Educational Psychology, 10*, 342-348.

Lea, M., & Spears, R. (1991). Computer-mediated communication, de-individuation and group decision-making. *International Journal of Man-Machine Studies, 34*, 283-301.

Leitão, S. (2000). The potential of argument in knowledge building. *Human Development, 43*, 332-360.

Leitão, S., & Almeida, E. G. (2000). A producão de contra-argumentos na escrita infantil [Counterargument in children's writing]. *Psicologia: Reflexão e Crítica, 13*(3), 351-361.

Linn, M., & Burbules, N. C. (1993). Construction of knowledge and group learning. In K. Tobin (Ed.), *The practice of constructivism in science education* (pp. 91-119). Washington, DC: American Association for the Advancement of Science (AAAS).

Littleton, K., & Häkkinen, P. (1999). Learning together: Understanding the processes of computer-based collaborative learning. In P. Dillenbourg (Ed.), *Collaborative learning: Cognitive and computational approaches* (pp. 20-30). Amsterdam: Pergamon.

Mandl, H., & Fischer, F. (Eds.). (2000). *Wissen sichtbar machen [Making knowledge visisble]*. Göttingen: Hogrefe.

Mandl, H., & Fischer, F. (in press). Computer networking for education. In N. J. Smelser & P. B. Baltes (Eds.), *International Encyclopedia of the Social & Behavioral Sciences*. Amsterdam: Pergamon.

Mandl, H., & Friedrich, H. F. (Eds.). (1992). *Lern- und Denkstrategien. Analyse und Intervention [Strategies of learning and thinking. Analysis and intervention]*. Göttingen: Hogrefe.

Mandl, H., Gruber, H., & Renkl, A. (1991). Kontextualisierung von Expertise [Contextualization of expertise]. In H. Mandl, M. Dreher & H.-J. Kornadt (Eds.), *Entwicklung und Denken im kulturellen Kontext [Development and thinking in a cultural context]* (pp. 203-227). Göttingen: Hogrefe.

Mandl, H., Gruber, H., & Renkl, A. (1993). Misconceptions and knowledge compartmentalization. In G. Strube & F. Wender (Eds.), *The cognitive psychology of knowledge: The German Wissenspsychologie project* (pp. 161-176). Amsterdam: Elsevier.

Mandl, H., Gruber, H., & Renkl, A. (1994a). Knowledge application in complex situations. In S. Vosniadou, E. De Corte & H. Mandl (Eds.),

*Technology-based learning environments. Psychological and educational foundations* (pp. 40-47). Berlin: Springer.

Mandl, H., Gruber, H., & Renkl, A. (1994b). Zum Problem der Wissensanwendung [The problem of knowledge application]. *Unterrichtswissenschaft, 22*(3), 233-242.

Mandl, H., Gruber, H., & Renkl, A. (1996). Communities of practice toward expertise: Social foundation of university instruction. In P. B. Baltes & U. Staudinger (Eds.), *Interactive minds. Life-span perspectives on the social foundation of cognition* (pp. 394-411). Cambridge: Cambridge University Press.

Mandl, H., & Renkl, A. (1992). A Plea for "More Local" Theories of Cooperative Learning. *Learning and Instruction, 2*(1992), 281 - 285.

Markus, M. L. (1987). Towards a "critical mass" theory of interactive media: Universal access, interdependence, and diffusion. *Communication Research, 14*, 491-511.

Marttunen, M., & Laurinen, L. (2001). Learning of argumentation skills in networked and face-to-face environments. *Instructional Science, 29*, 127-153.

Mason, L. (1998). Sharing cognition to construct scientific knowledge in school context: The role of oral and written discourse. *Instructional Science, 26*, 359-389.

Matheson, K., & Zanna, M. P. (1990). Computer-mediated communications: The focus is on me. *Social Science Computer Review, 8*(1), 1-12.

McGrath, J. E., & Hollingshead, A. B. (1993). Putting the "group" back in group support systems: Some theoretical issues about dynamic processes in groups with technological enhancements. In L. M. Jessup & J. S. Valacich (Eds.), *Group support systems: New perspectives* (pp. 78-96). New York: Macmillan.

McGrath, J. E., & Hollingshead, A. B. (1994). *Groups interacting with technology: Ideas, evidence, issues, and an agenda.* Thousand Oaks, CA: Sage.

McLaughlin, M. W. (1976). Implementation as mutual adaption: Change in classroom organisation. *Teachers College Record, 77*(3), 339-351.

Means, M. L., & Voss, J. F. (1996). Who reasons well? Two studies of informal reasoning among children of different grade, ability and kowledge levels. *Cognition and Instruction, 14*(139-178).

Mercer, N. (1995). *The guided construction of knowledge. Talk among teachers and learners.* Clevedon: Multilingual Matters.

Mercer, N., & Wegerif, R. (1999). Is 'exploratory talk' productive talk? In K. Littleton & P. Light (Eds.), *Learning with computers. Analysing productive interaction* (pp. 79-101). London: Routledge.

Mok, M. (1996). *Sample size requirements for 2-level designs in educational research.* London: Multilevel Models Project.

Nastasi, B. K., & Clements, D. H. (1991). Research on cooperative learning: Implications for practice. *School Psychology Review, 20*, 110-131.

Nastasi, B. K., & Clements, D. H. (1992). Social-cognitive behaviors and higher-order thinking in educational computer environments. *Learning and Instruction, 2*, 215-238.

Naveh-Benjamin, M. (1991). A comparison of training programs intended for different types of test-anxious students: Further support for an information-processing model. *Journal of Educational Psychology, 83*, 134-139.

Neuman, Y., Leibowitz, L., & Schwarz, B. B. (2000). Patterns of verbal mediation during problem solving: A sequential analysis of self-explanation. *The Journal of Experimental Education, 68*(3), 197-213.

Ng, E., & Bereiter, C. (1991). Three levels of goal orientation in learning. *The Journal of the Learning Sciences, 1*(3&4), 243-271.

Nussbaum, E. M., Hartley, K., Sinatra, G. M., Reynolds, R. E., & Bendixen, L. D. (2002, April). *Enhancing the quality of on-line discussions.* Paper presented at the Annual meeting of the American Educational Research Association, New Orleans, LA.

O'Donnell, A. M. (1999). Structuring dyadic interaction through scripted cooperation. In A. M. O'Donnell & A. King (Eds.), *Cognitive perspectives on peer learning* (pp. 179-196). Mahwah, NJ: Erlbaum.

O'Donnell, A. M., & Dansereau, D. F. (1992). Scripted cooperation in student dyads: A method for analyzing and enhancing academic learning and performance. In R. Hertz-Lazarowitz & N. Miller (Eds.), *Interactions in cooperative groups. The theoretical anatomy of group learning* (pp. 120-141). Cambridge, MA: Cambridge University Press.

O'Donnell, A. M., & Dansereau, D. F. (2000). Interaction effects of prior knowledge and material format on cooperative teaching. *Journal of Experimental Education, 68*, 101-118.

O'Donnell, A. M., Dansereau, D. F., Hall, R. H., & Rocklin, T. R. (1987). Cognitive, social/affective, and metacognitive outcomes of scripted cooperative learning. *Journal of Educational Psychology, 79*(4), 431-437.

O'Donnell, A. M., & King, A. (Eds.). (1999). *Cognitive perspectives on peer learning.* Mahwah, NJ: Erlbaum.

Paivio, A. (1986). *Mental representations. A dual coding approach.* New York: Oxford University Press.

Palincsar, A. S., & Brown, A. L. (1984). Reciprocal teaching of comprehension-fostering and monitoring activities. *Cognition and Instruction, 1*, 117-175.

Palincsar, A. S., & Herrenkohl, L. R. (1999). Designing collaborative contexts: Lessons from three research programs. In A. M. O'Donnell &

A. King (Eds.), *Cognitive perspectives on peer learning* (pp. 151-177). Mahwah, NJ: Erlbaum.

Pea, R. D. (1994). Seeing what we build together: Distributed multimedia learning environments for transformative communications. Special Issue: Computer support for collaborative learning. *Journal of the Learning Sciences, 3*(3), 285-299.

Person, N. K., & Graesser, A. G. (1999). Evolution of discourse during cross-age tutoring. In A. M. O'Donnell & A. King (Eds.), *Cognitive Perspectives on Peer Learning* (pp. 69-86). Mahwah, NJ: Erlbaum.

Pfister, H.-R., & Mühlpfordt, M. (2002). *Supporting discourse in a synchronous learning environment: The learning protocol approach.* Paper presented at the Conference on Computer Support for Collaborative Learning (CSCL), Boulder, CO.

Piaget, J. (1928). *Judgement and reasoning in the child.* London: Routledge.

Piaget, J. (1985). *The equilibrium of cognitive structures: The central problem of intellectual development.* Chicago: University of Chicago Press.

Pontecorvo, C., & Girardet, H. (1993). Arguing and reasoning in understanding historical topics. *Cognition and Instruction, 11 (3&4)*, 365-395.

Prenzel, M., Eitel, F., Holzbach, R., Schoenhein, R.-J., & Schweiberer, L. (1993). Lernmotivation im studentischen Unterricht in der Chirurgie [Students' learning motivation in surgery instruction]. *Zeitschrift für Pädagogische Psychologie, 7*, 125-137.

Quinn, C. N., Mehan, H., Levin, J. A., & Black, S. D. (1983). Real education in non-real time: The use of electronic message systems for instruction. *Instructional Science, 4*, 313-327.

Reeve, J. (1996). *Motivating others: Nurturing inner motivational resources.* Boston, MA: Allyn & Bacon.

Reif, F., & Heller, J. I. (1982). Knowledge structures and problem solving in physics. *Educational Psychologist, 17*, 102-127.

Reinmann-Rothmeier, G. (2003). *Didaktische Innovation durch Blended Learning [Didactical innovation through blended learning]*. Bern: Huber.

Reinmann-Rothmeier, G., & Mandl, H. (2002). Analyse und Förderung kooperativen Lernens in netzbasierten Umgebungen [Analysis and facilitation of cooperative learning in net-based environments]. *Zeitschrift für Entwicklungspsychologie und Pädagogische Psychologie, 31*(1), 44-57.

Reinmann-Rothmeier, G., & Mandl, H. (Eds.). (2001). *Virtuelle Seminare in Hochschule und Weiterbildung [Virtual seminars in university and professional development]*. Bern: Huber.

Reiser, B. J. (2002). *Why scaffolding should sometimes make tasks more difficult for learners*. Paper presented at the Computer Support for Collaborative Learning: Foundations for a CSCL Community, Boulder, CO.

Reiserer, M. (2003). *Peer-Teaching in Videokonferenzen. Effekte niedrig und hoch strukturierter Kooperationsskripts auf Lerndiskurs und Lernerfolg [Peer-teaching through video conferencing. Effects of low- and high-structured cooperation scripts on learning discourse and learning success]*. Berlin: Logos.

Renkl, A. (1997). *Lernen durch Lehren - Zentrale Wirkmechanismen beim kooperativen Lernen [Learning through teaching - central mechanisms in cooperative learning]*. Wiesbaden: Deutscher Universitäts-Verlag.

Renkl, A., Gruber, H., & Mandl, H. (1996). Kooperatives problemorientiertes Lernen in der Hochschule [Cooperative problem-oriented learning at the university]. In J. Lompscher & H. Mandl (Eds.), *Lehr- und Lernprobleme im Studium. Bedingungen und Veränderungsmöglichkeiten [Teaching and learning problems in university studies. Conditions and possibilites for change]* (pp. 131-147). Bern: Huber.

Renkl, A., & Mandl, H. (1995). Kooperatives Lernen: Die Frage nach dem Notwendigen und dem Ersetzbaren [Cooperative learning: A question of what is necessary and replaceable]. *Unterrichtswissenschaft, 23,* 292-300.

Resnick, L. B. (1987). Learning in school and out. *Educational Researcher, 16,* 13-20.

Rice, R. E. (1984). Mediated group communication. In R. E. Rice (Ed.), *The new media: Communication, research, and technology* (pp. 129-156). Beverly Hills, CA: Sage.

Rice, R. E. (1993). Media appropriateness. Using social presence theory to compare traditional and new organizational media. *Communication Research, 19*(4), 451-484.

Richter, T., Naumann, J., & Groeben, N. (2001). Das Inventar zur Computerbildung (INCOBI): Ein Instrument zur Erfassung von Computer Literacy und computerbezogenen Einstellungen bei Studierenden der Geistes- und Sozialwissenschaften [Inventory of computer education. A tool for surveying computer literacy and computer-related attitudes of students in humanities and social sciences]. *Psychologie in Erziehung und Unterricht, 48,* 1-13.

Riel, M. (1996). Cross-classroom collaboration: Communication and education. In T. Koschmann (Ed.), *CSCL: Theory and practice of an emerging paradigm* (pp. 187-207). Mahwah, NJ: Erlbaum.

Roschelle, J. (1996). Learning by collaborating: Convergent conceptual change. In T. Koschmann (Ed.), *CSCL: Theory and practice of an emerging paradigm* (pp. 209-248). Mahwah, NJ: Erlbaum.

Roschelle, J., & Pea, R. (1999). Trajectories from today's WWW to a powerful educational infrastructure. *Educational Researcher, 28*(5), 22-25 + 43.

Roschelle, J., & Teasley, S. D. (1995). The construction of shared knowledge in collaborative problem solving. In C. O'Malley (Ed.), *Com-*

*puter supported collaborative learning* (Vol. 128, pp. 69-97). Berlin: Springer.

Rosenshine, B., & Meister, C. (1994). Reciprocal teaching: A review of the research. *Review of Educational Research, 64*, 479-630.

Rosenshine, B., Meister, C., & Chapman, S. (1996). Teaching students to generate questions: A review of the intervention studies. *Review of Educational Research, 66*(2), 181-221.

Rost, D. H., & Schermer, F. J. (1997). *Differentielles Leistungsangst Inventar (DAI) [Differential inventory of performance anxiety (DAI)].* Frankfurt: Swets Test Services.

Rummel, N., & Spada, H. (2005). Learning to collaborate. An instructional approach to promoting collaborative problem-solving in computer-mediated settings. *Journal of the Learning Sciences, 14*(2), 201-241.

Salomon, G. (1993a). No distribution without individuals cognition: A dynamic interactional view. In G. Salomon (Ed.), *Distributed cognitions: Psychological and educational considerations.* Cambridge: Cambridge University Press.

Salomon, G. (Ed.). (1993b). *Distributed cognitions: Psychological and educational considerations.* Cambridge: Cambridge University Press.

Salomon, G., & Globerson, T. (1989). When teams do not function the way they ought to. *International Journal of Educational Research, 13*(1), 89-99.

Salomon, G., & Perkins, D. N. (1998). Individual and social aspects of learning. *Review of Research in Education, 23*, 1-24.

Scanlon, E., Issroff, K., & Murphy, P. (1999). Collaborations in a primary classroom. Mediating science activities through new technology. In K. Littleton & P. Light (Eds.), *Learning with computers. Analysing productive interaction* (pp. 62-78). London: Routledge.

Scardamalia, M., & Bereiter, C. (1994). Computer support for knowledge-building communities. *Journal of the Learning Sciences, 3*(3), 265-283.

Scardamalia, M., & Bereiter, C. (1996). Computer support for knowledge-building communities. In T. Koschmann (Ed.), *CSCL: Theory and practice of an emerging paradigm* (pp. 249-268). Mahwah, NJ: Erlbaum.

Scardamalia, M., Bereiter, C., McLean, R. S., J., S., & Woodruff, E. (1989). Computer supported intentional learning environments. *Journal of Educational Computing Research, 5*, 51-68.

Schank, R. C., & Abelson, R. P. (1977). *Scripts, plans, goals and understanding. An inquiry into human knowledge structures.* Hillsdale, NJ: Erlbaum.

Schmidt, H. G. (1983). Problem-based learning. Rationale and description. *Medical Education, 17*, 11-16.

Schmitz, J., & Fulk, J. (1991). Organizational colleagues, media richness, and electronic mail. A test of the social influence model of technology use. *Communication Research, 18*(4), 487-523.

Schnotz, W., Boeckheler, J., & Grzondziel, H. (1997). *Individual and co-operative acquisition of knowledge with static and animated pictures in computer-based learning environments.* Paper presented at the EARLI '97, Athen.

Schoenfeld, A. H. (1985). *Mathematical problem-solving.* San Diego, CA: Academic Press.

Schwarz, B. B., Neuman, Y., & Biezuner, S. (2000). Two wrongs may make a right ... if they argue together! *Cognition and Instruction, 18*(4), 461-494.

Schweizer, K., Paechter, M., & Weidenmann, B. (2001). A field study on distance education and communication: Experiences of a virtual tutor. *Journal of Computer-Mediated Communication, 6*(2). Retreived April 15, 2003, from http://www.ascusc.org/jcmc/vol6/issue2/schweizer.html

Seipp, B., & Schwarzer, C. (1991). Angst und Leistung - Eine Meta-Analyse empirischer Befunde [Anxiety and achievement - a meta-analysis of empirical findings]. *Zeitschrift für Pädagogische Psychologie, 5*, 85-97.

Sellin, N. (1990). On aggregation bias. *International Journal of Educational Research, 14*, 257-268.

Simon, H. A. (1955). A behavioral model of rational choice. *Quarterly Journal of Economics, 69*, 99-118.

Slavin, R. E. (1992). When and why does cooperative learning increase achievement? Theoretical and empirical perspectives. In R. Hertz-Lazarowitz & N. Miller (Eds.), *Interaction in cooperative groups. The theoretical anatomy of group learning* (pp. 145-173). Cambridge: Cambridge University Press.

Slavin, R. E. (1993). Synthesis of research on cooperative learning. In A. E. Woolfolk (Ed.), *Readings & cases in educational psychology* (pp. 170-178). Needham Heights: Allyn & Bacon.

Slavin, R. E. (1996). Research for the future. Research on cooperative learning and achievement: What we know, what we need to know. *Contemporary Educational Psychology, 21*, 43-69.

Snow, R. E. (1989). Toward assessment of cognitive and conative structures in learning. *Educational Researcher, 18*, 8-15.

Sorrentino, R. M., Short, J. C., & Raynor, J. O. (1984). Uncertainty orientation: Implications for affective and cognitive views of achievement behaviour. *Journal of Personality and Social Psychology, 46*, 189-206.

Spears, R., Lea, M., & Lee, S. (1990). De-individuation and group polarization in computer-mediated communication. *British Journal of Social Psychology, 29*, 121-134.

Spiro, R. J., Feltovich, P. J., Coulson, R. L., & Anderson, D. K. (1989). Multiple analogies for complex concepts: Antidotes for analogy-induced misconceptions in advanced knowledge acquisition. In S. Vosniadou & A. Ortony (Eds.), *Similarity and analogical reasoning* (pp. 498-531). Cambridge: Cambridge University Press.

Stark, R., Gruber, H., Renkl, A., & Mandl, H. (1997). "Wenn um mich herum alles drunter und drüber geht, fühle ich mich so richtig wohl" - Ambiguitätstoleranz und Transfererfolg ["If everything is topsy-turvy, I feel at home" - tolerance of ambiguitiy and transfer]. *Psychologie in Erziehung und Unterricht, 44*, 204-215.

Stark, R., & Mandl, H. (2002, September). *"Unauffällige", "Vorwissensschwache", "Unmotivierte" und "Musterschüler": Homogene Untergruppen beim Lernen mit einem komplexen Lösungsbeispiel im Bereich empirischer Forschungsmethoden ["Inconspicious," "less-able," "unmotivated," and "model students": Homogeneous sub-groups in learning with complex worked out examples in the domain of empirical research methods].* Research report 147, Ludwig Maximilian University of Munich, Institute of Educational Psychology, presented at the 43rd congress of the Deutsche Gesellschaft für Psychologie, Berlin/Germany.

Stegmann, K. (2002). *NetBite – ein virtuelles Tutorium für die Empirischen Forschungsmethoden in der Pädagogik. Konzeption & Evaluation [NetBite - a virtual tutorial for empirical research methods in pedagogy. Conceptualization & evaluation].* Unpublished master thesis, Ludwig Maximilian University of Munich.

Stegmann, K., Weinberger, A., & Fischer, F. (2007). Facilitating argumentative knowledge construction with computer-supported collaboration scripts. *International Journal of Computer-Supported Collaborative Learning, 2*(4), 421-447.

Stonewater, J. K. (1980). Strategies for problem solving. In R. E. Young (Ed.), *Fostering critical thinking* (Vol. 3, pp. 33-57). San Francisco, CA: Jossey-Bass.

Straus, S. G., & McGrath, J. E. (1994). Does the medium matter? The interaction of task type and technology on group performance and member reactions. *Journal of Applied Psychology, 79*(1), 87-97.

Teasley, S. (1997). Talking about reasoning: How important is the peer in peer collaboration? In L. B. Resnick, R. Säljö, C. Pontecorvo & B. Burge (Eds.), *Discourse, tools and reasoning: Essays on situated cognition* (pp. 361-384). Berlin: Springer.

Teasley, S. D., & Roschelle, J. (1993). Constructing a joint problem space: The computer as a tool for sharing knowledge. In S. P. Lajoie & S. J. Derry (Eds.), *Discourse, tools, and reasoning: Essays on situated cognition* (pp. 229-258). Berlin: Springer.

Tudge, J. (1989). When collaboration leads to regression: Some negative consequences of socio-cognitive conflict. *European Journal of Social Psychology, 19*, 123-138.

Tudge, J. R. H. (1992). Processes and consequences of peer collaboration: A Vygotskian analysis. *Child Development, 63*(1364-1379).

Veerman, A. L., & Treasure-Jones, T. (1999). Software for problem solving through collaborative argumentation. In P. Coirier & J. E. B. Andriessen (Eds.), *Foundations of argumentative text processing* (pp. 203-230). Amsterdam: Amsterdam University Press.

Voss, J. F., Greene, T. R., Post, T. A., & Penner, B. C. (1983). Problem-solving skills in the social sciences. *The Psychology of Learning and Motivation, 17*, 165-213.

Vye, N. J., Goldman, S. R., Voss, J. F., Hmelo, C., Williams, S., & Cognition and Technology Group at Vanderbilt. (1997). Complex mathematical problem solving by individuals and dyads. *Cognition and Instruction, 15*(4), 435-484.

Vygotsky, L. S. (1978). *Mind in society. The development of higher psychological processes*. Cambridge: Harvard University Press.

Vygotsky, L. S. (1986). *Thought and language* (2nd ed.). Cambridge, MA: MIT Press.

Walther, J. B. (1992). Interpersonal effects in computer-mediated interaction. A relational perspective. *Communication Research, 19*(1), 52-90.

Walther, J. B. (1996). Computer-mediated communication: Impersonal, interpersonal, and hyperpersonal interaction. *Communication Research, 23*(1), 3-43.

Walton, D. N., & Krabbe, E. C. W. (1995). *Commitment in dialogue. Basic concepts of interpersonal reasoning.* Albany, NY: State University of New York Press.

Webb, N. M. (1989). Peer interaction and learning in small groups. *International Journal of Educational Research, 13*, 21-39.

Webb, N. M. (1991). Task-related verbal interaction and mathematics learning in small groups. *Journal of Research in Mathematics Education, 22*, 366-389.

Webb, N. M. (1992). Testing a theoretical model of student interaction and learning in small groups. In R. Hertz-Lazarowitz & N. Miller (Eds.), *Interaction in cooperative groups:The theoretical anatomy of group learning* (pp. 102-119). New York: Cambridge University Press.

Webb, N. M., Ender, P., & Lewis, S. (1986). Problem-solving strategies and group processes in small groups learning computer programming. *American Educational Research Journal, 23*(2), 243-261.

Webb, N. M., & Farivar, S. (1994). Promoting helping behavior in cooperative small groups in middle school mathematics. *American Educational Research Journal, 31*, 369-395.

Webb, N. M., Jonathan, D., Fall, T., & Fall, R. (1995). Constructive actvity and learning in collaborative small groups. *Journal of Educational Psychology, 87*, 406-423.

Weinberger, A. (1998). *Grounding in kooperativen Lernumgebungen: Empirische Untersuchung von Aushandlungsprozessen in computervermittelter und Face-to-Face-Kommunikation [Grounding in cooperative learning environments: Empirical study of negotiation processes in computer-mediated and face-to-face communication].* Unpublished master thesis, Ludwig Maximilian University of Munich.

Weinberger, A. & Fischer, F. (2006). A framework to analyze argumentative knowledge construction in computer-supported collaborative learning. *Computers & Education, 46*(1), 71-95.

Weinberger, A., Fischer, F., & Mandl, H. (2002). *Kodierungssystem für eine Multi-Ebenen-Analyse der gemeinsamen Wissenskonstruktion [Coding system for a mult-level-analysis of collaborative knowledge construction].* Unpublished manuscript, Ludwig Maximilian University of Munich, Institute of Educational Psychology.

Weinberger, A., & Lerche, T. (2001). Didaktik der neuen Informations- und Kommunikationstechnologien [Didactics of new information and communication technologies]. In A. Hug (Ed.), *Wie kommt Wissenschaft zu ihrem Wissen? [How does science acquire knowledge?]* (Vol. 1, pp. 359-375). Baltmannsweiler Hohengehren: Schneider.

Weinberger, A., & Mandl, H. (2001). Wandel des Lernens durch Neue Medien - das virtuelle Seminar Empirische Erhebungs- und Auswertungsverfahren [Changing learning through new media - the virtual seminar "Methods of empirical inquiry and analysis"]. In H. F. Friedrich & F. Hesse (Eds.), *Partizipation und Interaktion im virtuellen Seminar [Participation and interaction in the virtual seminar]* (pp. 243-268). Münster: Waxmann.

Weinberger, A., & Mandl, H. (2003). Computer-mediated knowledge communication. Special Issue: New Media in Education. *Studies in Communication Sciences*, 81-105.

Weinberger, A., Reiserer, M., Ertl, B., Fischer, F., & Mandl, H. (2005). Facilitating collaborative knowledge construction in computer-mediated learning with structuring tools. In R. Bromme, F. Hesse & H.

Spada (Eds.), *Barriers and biases in computer-mediated knowledge communication - and how they may be overcome* (pp. 15-37). Boston: Kluwer.

Weinberger, A., Stegmann, K., Fischer, F., & Mandl, H. (2003). *Kooperatives Lernen im Netz: Förderung der argumentativen Wissenskonstruktion mit Kooperationsskripts [Cooperative learning in the internet: Facilitation of argumentative knowledge construction with cooperation scripts].* Paper presented at the Netzbasierte Wissenskommunikation in Gruppen, Tübingen, Germany.

Weinberger, A., Stegmann, K., & Fischer, F. (2007). Knowledge convergence in collaborative learning: Concepts and assessment. *Learning & Instruction, 17*(4), 416-426.

Weiner, B. (1985). An attributional theory of achievement motivation and emotion. *Psychological Review, 92*, 548-573.

Weiner, B., Frieze, I., Kukla, A., Reed, L., Rest, S., & Rosenbaum, R. M. (1971). *Perceiving the causes of success and failure.* Morristown, NJ: General Learning Process.

Weinert, F. E. (1994). Lernen lernen und das eigene Lernen verstehen [Learning to learn and understanding one's learning]. In K. Reusser & M. Reusser-Weyeneth (Eds.), *Verstehen. Psychologischer Prozeß und didaktische Aufgabe [Understanding. Psychological processes and didactical tasks]* (pp. 183-205). Bern: Huber.

Wild, K.-P., & Schiefele, U. (1994). Lernstrategien im Studium: Ergebnisse zur Faktorenstruktur und Reliabilität eines neuen Fragebogens [Learning strategies in university studies: Factor structure and reliability of a new questionnaire]. *Zeitschrift für Differentielle und Diagnostische Psychologie, 15*, 185-200.

Wiley, J., & Voss, J. F. (1999). Constructing arguments from multiple sources: Tasks that promote understanding and not just memory for text. *Journal of Educational Psychology, 91*(2), 301-311.

Winkler, K., & Mandl, H. (2002). Knowledge Master: Wissensmanagement-Weiterbildung durch E-Learning [Knowledge Master: Professional development through e-learning]. In U. Dittler (Ed.), *E-Learning - Einsatzkonzepte und Erfolgsfaktoren des computergestützten und multimedialen Lernens mit interaktiven Medien [E-learning - concepts and factors of computer-supported and multi-media learning with interactive media]* (pp. 205-215). München: Oldenbourg.

Woodruff, E. (1995). *The effects of computer mediated communications on collaborative discourse in knowledge-building communities.* Paper presented at the Annual meeting of the American Educational Research Association, San Francisco.

Yale Corporation. (1828). *Reports on the course of instruction.*Unpublished manuscript, New Haven.

Yin, R. K. (2003). *Case study research: Design and methods* (3rd ed.). Thousand Oaks, CA: Sage.

Ziegler, A., & Heller, K. (2000). Effects of an attribution retraining with female students gifted in physics. *Journal for the Education of the Gifted, 23*, 217-243.